MW00836908

Optoelectronic Packaging

WILEY SERIES IN MICROWAVE AND OPTICAL ENGINEERING

KAI CHANG, Editor
Texas A&M University

Optoelectronic Packaging

Edited by

ALAN R. MICKELSON
Department of Electrical & Computer Engineering
University of Colorado at Boulder

NAGESH R. BASAVANHALLY
Lucent Technologies
Bell Laboratories

YUNG-CHENG LEE
Department of Mechanical Engineering
University of Colorado at Boulder

A WILEY-INTERSCIENCE PUBLICATION

JOHN WILEY & SONS, INC.

NEW YORK / CHICHESTER / WEINHEIM / BRISBANE / SINGAPORE / TORONTO

Library of Congress Cataloging in Publication Data:

Optoelectronic packaging / edited by A. R. Mickelson, N. R.
   Basavanhally, Y. C. Lee.
      p. cm. – (Wiley series in microwave and optical engineering)
   Includes bibliographical references.
   ISBN 0-471-11188-0 (cloth : alk. paper)
   1. Optoelectronic devices. 2. Microelectronic packaging.
  3. Integrated optics. I. Mickelson, Alan Rolf, 1950-
  II. Basavanhally, N. R. (Nagesh R.) III. Lee, Yung-Cheng.
  IV. Series.
  TA1750.059   1997
  621.381'045–dc20   96-44716

Contributors

Nagesh R. Basavanhally, Lucent Technologies, Princeton, NJ

Venkata A. Bhagavatula, Corning, Inc., Corning, NY

Mario Dagenais, University of Maryland, College Park, MD

Robert F. Feuerstein, Optoelectronic Computing Systems Center, University of Colorado, Boulder, CO

S. Fox, University of Maryland, College Park, MD

Gary J. Grimes, Center for Telecommunications Education & Research, University of Alabama, Birmingham, AL

Paul Haugsjaa, GTE Laboratories, Inc., Waltham, MA

Robert Holland, Lucent Technologies, Princeton, NJ

Felix Kapron, Bellcore, Morristown, NJ

Y. C. Lee, Department of Mechanical Engineering, University of Colorado, Boulder, CO

Wei Lin, Department of Mechanical Engineering, University of Colorado, Boulder, CO

Timothy McLaren, Department of Mechanical Engineering, University of Colorado, Boulder, CO

Scott A. Merritt, University of Maryland, College Park, MD

Alan R. Mickelson, Department of Electrical & Computer Engineering, University of Colorado, Boulder, CO

K. Mobarhan, University of Maryland, College Park, MD

John Neff, Optoelectronic Computing Systems Center, University of Colorado, Boulder, CO

Ronald Nordin, Lucent Technologies, Naperville, IL

Laurence Watkins, Lucent Technologies, Princeton, NJ

R. Whaley, University of Maryland, College Park, MD

Naoaki Yamanaka, Nippon Telegraph and Telephone Corporation, Tokyo, Japan

Contents

Preface

It is only recently that packaging has emerged as a field in itself. The complexity of electronic circuits really demanded this in the case of electronic packaging. As chip pin counts increased, the interconnection problem became ever more complex. It was no longer feasible to simply tie pins together; instead, thought had to be given to layout, which led to more thought about electrical noise, thermal issues, and mechanical issues. In the case of optoelectronics, a comparable packaging situation emerged from day one. The original semiconductor lasers ran very hot, yet heatsinking had to be achieved without significant sacrifice of coupling efficiency. Lasers may run cooler today, but new problems have replaced the old. The state of the art requires coupling to single-mode waveguides requiring submicron alignment tolerances. To compete with electronics, there are those who would claim that one must use optical parallelism. Although single-mode lasers can be made to run cool, one will always run an array with as many elements as thermal tolerance will allow. Laser array coupling, therefore, requires both state-of-the-art optomechanical as well as thermal packaging—and this ignores the electrical crosstalk problems associated with modulation problems, electrical packaging problems as daunting as the most complex from electronic packaging. Optoelectronic packaging issues compound all of those seen to the present in other fields with a host of new ones.

The present volume is a collection of contributions that is meant to provide an overview of the techniques being used or contemplated for packaging of optoelectronic devices. The authors are all practitioners of packaging science, and this fact in itself imparts a sense of reality to the contents. In this volume, one will find details of the nitty gritty of existing commercial packages as well as some blueprints for soon-to-be packages and even some more blue sky considerations relating to the more futuristic "what if" types of packages.

The purpose of the volume is multifold. Although the level of the material is introductory, the packaging expert can use the contents to determine what techniques are being pursued at what location. The expert from another field can use the contents to try to determine what the present-day state of the art of optoelectronics is as well as to determine where the technology may go. For the neophyte, the volume can serve as a definition of what is meant by packaging or, more specifically, optoelectronic packaging, while serving as an introduction to the more detailed literature.

The scope of the book is really limited to assembly technologies. The topic of packaging materials would require a volume in itself. The inclusion of such material was in the original plan, but the quantity of material on assembly that we have received has precluded the inclusion of more material. Perhaps there can be a later volume on optoelectronic packaging materials. There is also a limited coverage of systems-oriented packaging design for manufacturability, although the volume is by no means empty of design considerations.

There really is no comparable volume on the market. This tome contains significant material that does not appear elsewhere, including a number of case studies of packaged subassemblies. We feel that the volume can serve as a useful tool to a large audience. We feel that, as optoelectronics increases in importance as an adjunct technology to electronics, the audience that could profit from a perusal of this volume should increase continuously.

The editors sincerely thank all the authors who contributed invaluable manuscripts to this book. We are indebted to Ms. Julie Fredlund for her persistence and meticulous help in editorial work.

University of Colorado at Boulder   A. R. MICKELSON
Bell Laboratories   N. R. BASAVANHALLY
University of Colorado at Boulder   Y. C. LEE

Optoelectronic Packaging

CHAPTER ONE

Introduction

ALAN R. MICKELSON

The present volume is a compendium of chapters by a number of authors from academia and industry describing various aspects of the field of optoelectronic packaging. To introduce this material, we will present a short chronology of electronic packaging, followed by a section discussing what is really meant by packaging before giving a brief overview of the chapters included in this volume.

1.1 A BRIEF HISTORY OF ELECTRONIC PACKAGING

The advent of the integrated circuit led to a new set of problems in the electronics industry, namely, those of mounting components. When components were all discrete with either two or three electrical leads emanating from their lumped bodies, there was no special topological problem involved with their interconnection. Leads could just be made longer and twisted and bent to pass over each other. There was no need for mounting to be even on boards, much less planarized on board-like surfaces.

It is hard to think of the integrated circuit as a problem in any real sense; it was more like a revolution—the first embodiment of the realization of the promise of the transistor. Only the packaging engineer could really lament. The integrated circuit suddenly put together a number of connections, greater than two or three, onto the same planar level. Furthermore, there was no need for wires to protrude from this planar layer. If necessary, some subset of the wiring could protrude (and indeed, until recently, packaging solutions have always in one way or another used some wire bond interconnects), but it was no longer necessary. Indeed, it was clear from day one of the IC Age that hanging spaghetti-like wires from the IC would all but negate the tremendous advantages of being able to integrate functions and achieve significantly higher processing speeds.

1

Until the more recent advent of flip-chip attachment, whereby a bare die (silicon chip with thin metallization but no attached wires) is flipped and directly soldered to a substrate, all packaging techniques have begun with a primary package. This primary package attached short, thin wires to the on-chip pads and flared them out. These thin wires were then hand-attached to a rigid mount in order to protect them. The mount generally had pins protruding that could then be attached to some form of board. Advances in the chemistry of polymers allowed these materials to be formed into boards and moldings which for both primary carrier and board-level mounts became ubiquitous.

At the board level, a still-common solution is the polymer laminate board, especially for problems where multiple-layer interconnects are important. In such a solution, polymers are molded into sheets and the sheets rolled off presses, processed with patterned metallizations on either side, and then combined under pressure to form a multilayer board. Via holes between the layers can be fabricated either before or after the lamination procedure. In the original through-the-board technology, all the interconnections were etched into planarized metal on the back of the board. Only simple interconnections can, at best, be implemented with such a single-layer interconnect technology, simply due to the basic rules of topology. The multilayer solution is a necessity for any board with any level of processing built in, such as for a personal computer (PC). Special-purpose military analog circuitry requires dimensional stability (i.e., polymers expand and contract with temperature at a rate which is significantly higher than that of certain other materials), and therefore the boards are often implemented in multilayer ceramic. Polymer, however, is the most cost-effective implementation technology. The ceramic implementation can cost many times as much as the polymer implementations simply due to processing considerations.

Reduction of cost has been the primary driving force behind electronic development for several years now. Miniaturization has been one of the techniques applied to achieve reduced costs. A route to miniaturization is to increase complexity at a fixed size. Increased complexity can be achieved through increasing density or increasing speed. These two routes are in general both necessary but are not always necessarily compatible, at least within a given packaging technology. IC transistor counts are ever increasing, with an increased pin count and, therefore, interconnect density. Increased speed, however, even at fixed interconnect spacing, will increase mutual interference between dies; that is, increased speed (i.e., clock rate, which dictates the maximum frequency transmitted) decreases wavelength, and the effective shielding of a line is the distance to the nearest ground plane in wavelengths. (We will discuss electromagnetic speed limitations further in a later introductory paragraph.) Transmission lines carry their ground planes along with them to fix this shielding distance and make it less than the distance to the nearest "hot" core. This can't be done at the pinout of a chip, at least not economically. One has to live with some level of crosstalk. A way to minimize this crosstalk is to simply make all leads as short as possible and completely eliminate loops and bends and twists, and so on. From this challenge developed a plethora of direct-attach-type technologies. A bare die will have a slew of bonding pads on top of the chip. In the microwave analog world, the number could be small, and careful wire bonding from the top to pads on a board can actually be a viable solution to the connection problem. In a world in which there is a significant amount of circuitry dedicated to processing of parallel high-speed signals or even more so in the everyday digital world, this density can be huge. The wire bonds will talk. Therefore, there are a growing number of technologies (as well as names given to these techniques) in which the chip is attached upside down on the board, eliminating the primary chip carrier completely. This requires that the chip bonding pad be directly connected to some form of board attachment, which is then either flared out into lines that attach to the vias (e.g., ball grid array) or directly attached to vias [direct chip attach (DCA) or flip-chip soldering]. Especially with soldering techniques, there will be a processing step in which there is an elevated (from room temperature and perhaps even into hundreds of degrees Celsius) temperature step. Many of the typical laminate boards, which can be acrylate or epoxy, decompose when exposed to temperatures greater than roughly 100°C. A new class of materials, however, was synthesized in the late 1960s which has significantly better temperature stability—the polyimides. Although polyimides are known to take up some amount of water, glass transition temperatures over 300°C and possibly above 400°C counteract many of these slight disadvantages. At this point in time, polyimides exist in forms that can be rolled (i.e., Kapton) to form the so-called MCM-L as well as numerous forms that can be spun on to form the so-called MCM-D. Polyimide is presently the material of choice for low-cost yet high-performance packaging. Alternative materials such as benzocyclobutenes and polyquinoxilenes, however, hold promise for improving on polyimide drawbacks, although they presently remain distant alternatives. /

Optics has become a technology of today for telecommunications and is becoming one (at least one for the near future) for data communications. There is really very little or no thought given to alternative technologies for terrestrial long haul. When local distribution presents high overhead, then sometimes satellite technology becomes meshed with long-haul optical networking, but optical networking is at present always the backbone and workhorse. Although originally optics was perhaps oversold as being the technology which would replace not only interconnection wire but also electronics and would thereby be the only technology in data communications, the actual optical solutions now being implemented are actually evolutionary and involve first replacing wires and coaxes with fiber. Telecom solutions, if anything, were one of the great hindrances to optics becoming the dominant datacom solution. The telecommunications right-of-way and installation costs are so high that the components and subsystems of telecom systems are, in essence, free. The drivers for optics in the telephone system were capacity and long-term scalability of that capacity, clearly not cost. Telecom transceivers are prohibitively expensive, and any element used in the telecom transceiver solution is wrong-minded for a datacom solution. Ongoing present-day efforts are generating ever more cost-effective optical transceiver solutions, though, and thereby moving optics more and more decidedly into the local area network (LAN) and medium area network (MAN) arenas. At present, the internet is fueling an insatiable thirst for bandwidth, and the bandwidths consumable by the internet (hundreds of M/s to G/s) can only be satiated by optical solutions. /

Will optical solutions continue to work themselves down to smaller and smaller dimensions? For those of us with long-term intellectual investments in information transmission optics, yes would be the answer of choice. Maybe this answer is not so whimsical. As was mentioned above, polyimide is presently the material of choice for electronic packaging. Recently developed (in the last five years) fluorinated polyimides can also have very favorable optical characteristics such as good transparency and low scattering. Other experts in the field have argued in terms of impedance (impedance being the ratio of the voltage to the current at a given point in a circuit, in particular at input and output pins of chips) that, theoretically, optics can become a more viable technology for interconnects than for electronics, all the way down to the 100-μm interconnect length [Mil89], were the world a perfect place and integrated optics a perfect technology. Here, let's try another argument that is a bit less theoretical and with which I actually got agreement from an MCM manufacturer. What we have seen in recent years is a tremendous growth in the PC market. This is in great part due to the fact that PC power grows unabatedly while supercomputers and mainframes seem to have stagnated in terms of processing power. I would claim that a major underlying driver for these developments is an idea related to the concept of wavelength. Electrical interconnects are cheap, at least in the way they are implemented in PCs. In a PC interconnect, the voltage levels are fixed. One has output pins on a chip with fixed voltage and reasonably high current (low impedance) which drive lines hooked to chip inputs which are designed to operate with an input of fixed voltage and lower current (higher impedance) than the output pin. The inputs by nature need to have lower current than the outputs they are hooked to, because these outputs, even when not fanned out, must drive lossy lines over finite distances. There is, therefore, as was previously pointed out, a fundamental impedance mismatch. (Again, impedance is just voltage divided by current at a point.)

Impedance mismatch in the high-frequency (microwave) analog world is a bad idea. At high enough frequency, lines begin to take on an appreciable length compared to a wavelength, and a reflection generated at a coupling point from a line to a pin can come back to the source line connection point out of phase with the source to cancel a portion of the source signal. With a short enough line, this is nothing to worry about, because the reflections will always be in phase as long as the line is less than the wavelength by, say, 10. In the microwave world, this is seldom the case, and the source impedance must be matched to a line, which must be a transmission line, which must carry along its own ground plane at a fixed distance from the conductor, and this line must be judiciously connected to a closely matched receiver impedance. In the PC world, use of lines with accompanying ground planes would require radically different designs from those used at present and would lead to prohibitive cost. In the PC world, the interconnect line carries no ground and has no well-defined impedance, which is acceptable if it is less than a tenth of a wavelength for the highest frequency to be carried. If the clock rate is 100 MHz, this corresponds to a 3-meter wavelength. Interconnects less than 30 cm require no concept of matching. This is easy to achieve in a PC and possible in a workstation but hard in a mainframe. At a gigahertz, the tenth wavelength mark is 3 cm, and clearly this will be achievable in idealized MCMs on a board. At 10 GHz, 3 mm will be very hard to achieve. An interconnect limitation is looming. Optics does offer a solution here, because the impedance limitation can be cast as a separate problem, that of matching to a modulator at the one end and to a detector at the other. Although it is not clear that these problems are easily solvable, they are different problems than those of the line-driver electrical interconnections and, therefore, perhaps not fundamentally limited.

1.2 WHAT IS PACKAGING?

Much of my participation in the packaging area was primed by my involvement in a series of workshops that we (Y. C. Lee and myself) have put on primarily here in Colorado, a series which initiated in 1991 in Boulder and has, for the present at least, culminated in a 1994 workshop in Breckenridge. A problem plaguing any scientist or engineer who is trying to converse with a layperson about packaging is always terminology. "Packaging—isn't that what they do at Safeway?" "Gee, Table Mesa Shopping Center has a Packaging Center; is that related to your activity?" There was always the old New England term of "package store" which was applied to liquor stores, probably because some blue law disallowed use of even the term "liquor" in the zoning codes. In this light, there could be those who would consider the packaging center at the University of Colorado to be a place where refugees from the Ozarks congregate to design "stills." However, joking aside for the moment, I'm not sure it is clear either from the last section of this chapter or from all the voluminous hot air that emanated from the earlier of those workshops that anybody even slightly removed from the packaging area knows what the area is about any better than some of these laypeople.

The first of the Boulder planned packaging workshops was actually initiated by a year-end surplus in Al Harvey's (of the National Science Foundation's Lightwave Technology Division) conference workshop budget. Al thought that the Optoelectronic Computing Systems Center (OCSC) of the University of Colorado (CU) would be a natural place to hold a workshop which could discuss what NSF's role should be in funding the basic research areas that could contribute to advances in optoelectronic packaging techniques/technologies. The year of 1991 was the year that preceded a five-year downsizing trend which led to the abandonment of a number of industrial OE packaging efforts. The optoelectronic packaging area at that point in time was somewhat at a heyday, although with use of the term "heyday" there generally comes a connotation relating to the term "hype." The contents of the discussions at the workshop, as well as the report that emanated from it, had little to do with the initial purpose of the workshop and are somewhat indicative of the misunderstanding of the term "packaging." Although the NSF had in mind redirecting perhaps as much as hundreds of thousands of dollars in individual research contracts, discussion at the meeting rapidly turned to creating multiple hundreds-of-million-dollar regional centers. These centers would set up pilot lines to produce hundreds of already-demonstrated devices. There even seemed (not by a democratic vote, though, by any means) to be a consensus at the workshop that there was a gap in American technology: Perfectly good devices and subsystems that should be pilot-lined to allow manufacturing process control issues to be addressed and optical manufacturing methods to be developed were not being pursued because the development of the process technology was capital-intensive, and companies are loath to take on too many capital-intensive projects. There were a vocal minority at the workshop who thought that this should be the government's job. This was considered to be a major problem hindering scientific competitiveness in the United States. That this perceived problem may well be an actual problem I will not question. I don't think it has much if anything to do with packaging, however. I also think that it represented a misconception of the term "packaging" by a number of workshop attendees and led to considerable discussion that had little or nothing to do with the purpose of the workshop.

Quite generally, when a device physicist first fabricates an object or device that may do something new, that object is microscopic. One needs to have all kinds of wires and fibers and lenses, and so on, around it before he or she can find out what it can do. It's all of this junk around the microscopic device that is the package. Oftentimes, the first package is a mess. Generally, the first several efforts at making measurements of a new device/subsystem quite accurately characterize the package and have little or nothing to do with the device/subsystem. This is the stuff of packaging. Even advanced packaging relates to the creation of one or two devices, albeit maybe with enough forethought that measurements can be made that can de-embed the package characteristics from the device characteristics and, even further, that the package can be constructed such that it is manufacturing-compatible. This would mean that, although only one or two could be made by the techniques used to make the first one or two, there may exist techniques such that a manufacturing process could be created that could be optimized to make a large number of devices (whether cost-effectively or not is the subject matter of manufacturing engineering). It seems that it was this manufacturing process technology that was the stuff of the discussion at the first Boulder packaging meeting.

The second CU packaging workshop was held in late summer of 1992 at the Stanley Hotel in Estes Park, Colorado. The meeting was entitled The Symposium on Optoelectronic Packaging Science (dubbed SOEPS by my secretary Donald Hastings) and included time for discussion sessions in addition to the 15-minute talks which were presented by each of the 60-odd participants. One of the sessions, chaired by Professor Zoya Popović of CU, was entitled "Is Packaging a Science?," which was a theme of the workshop, along with an NSF-impressed theme (again, a portion of the funding came from Al Harvey, as is the case for all four) of "What Would a Packaging Curriculum Include?" Unfortunately, though, it seems that biases that arose at the first meeting recurred. It seems that at Zoya's session on packaging science the question was answered by decibel production rather than argument, and the resounding answer was that packaging is something that is done and has none of the elements of a science. This was considered to certainly preclude one from identifying a packaging curriculum. If packaging is simply a matter of mixing epoxies, methods developed to teach one to do this must be a waste of effort. At least at this point in time—that is, during the discussion in Zoya's session—I think that it was packaging that was being discussed, and those who had wanted the multimillion-dollar centers set up were now upset that talk would be given to the development of one or two devices. I feel, though, that the conclusion of that session was the last time during the four-workshop series that the manufacturing development point of view was espoused and that the last two workshops, as well as the contents of this book, will illustrate that packaging is a separate discipline which has all of the elements of a science.

The third and fourth workshops were held in Santa Barbara, California, and Breckenridge, Colorado, respectively. Both of these workshops were smaller—on the order of 30–40 attendees, and the attendees at these were probably all people involved in what I would define as packaging. The 1994 workshop in Breckenridge involved group problem solving, with the problem being that of design of a low-cost parallel optical link. Many of the attendees at that workshop are authors of this tome.

1.3   AN OVERVIEW OF THE CHAPTERS

The present book consists of a total of 12 contributed chapters, including this introductory chapter. As is the usual case, the original plan was a bit more elaborate than the resulting book, but the present volume seems to be of sufficient length and is reasonably uniform in content to warrant its standing alone as is. There had been an intention to include more chapters on packaging materials, but, as it turns out, such a topic could easily form a volume by itself. The material included in this volume, as was alluded to above, will cover the basics of packaging for optoelectronic interconnection, including some systems considerations in the earliest and latest chapters and including some of the more nitty gritty issues of laser, detector, waveguide, and passive device packaging in the middle.

Following this brief introduction, the development in this book then turns to systems considerations. The second chapter, by Naoaki Yamanaka of NTT Research, gives a rather general overview of the hierarchy of interconnections of communication systems (which clearly differs from the hierarchy of interconnections of computing systems so often discussed in optical computing circles) while including copious examples of NTT efforts to implement interconnection levels that are presently implemented with electronics. The third chapter, by Gary Grimes of AT&T and the University of Alabama at Huntsville, also has a systems flavor, although with an emphasis much more focused on the present-day telephone network. This chapter includes much information regarding the details of telecommunications electronics implementation which I have not seen appear elsewhere. Much of this implementation material is exactly the material lacking in courses on fiber-optic communications.

Chapter 4, by Felix Kapron of Bellcore and Robert Holland of AT&T*, covers the fundamentals of fiber and waveguide coupling and serves as a basis for some of the later material, especially that of Chapters 5 and 6. By nature, the optoelectronic packaging problem is one of interconnection of components, and it is useful to introduce this language of interconnection early in the volume. Chapter 5, written by Mario Dagenais' research group at the University of Maryland, covers the field of laser packaging. The chapter is notable for its attention to various details of such practical aspects as heatsinking, soldering, and facet coating. Chapter 6, by Laurence Watkins of AT&T, turns to detector packaging and includes some interesting examples of AT&T detector packages. Both Chapters 5 and 6 include a great amount of detail we have not seen elsewhere, and Chapter 6 also includes numerous examples of futuristic implementations. Chapter 7 is another background chapter, but it concentrates much more on waveguide characteristics than on coupling characteristics as did Chapter 4. Chapter 7 serves as a good introduction to Venkata Bhagavatula's Chapter 8 on passive device fabrication and packaging, which includes such practical considerations as passive techniques for pigtailing. The discussion of Corning's "holistic" approach to packaging (i.e., components are designed so as to give the best performance and cost when system-inserted rather than standing alone) is especially interesting. Examples of high-reliability construction connectors and applications of micro-V-groove technology also are included here that would be hard to find elsewhere.

The last four chapters of the volume turn to the slightly more futuristic area of parallel interconnections. Whether guided wave or free-space, many consider the real future of optical interconnects to lie in parallelism. The first of these four chapters, Chapter 9 by Nagesh Basavanhally of Lucent, Princeton and Ron Nordin of Lucent, Naperville, IL, reviews techniques of parallel waveguide-based device packaging, including a number of examples of ideas being worked out at Lucent in Princeton. Chapter 10, by Paul Haugsjaa of GTE Laboratories, reviews some novel ideas for array device packaging using silicon wafer boards for mounting of the optical waveguide devices and electronic devices together. Chapters 9 and 10, although discussing a more futuristic technology, discuss ways to achieve it that are "close in"—that is, in a major sense, incremental (or perhaps somewhat supraincremental) advancements over the kinds of already

---

* This division of AT&T is now Lucent Technologies.

available commercial processes discussed in Chapter 8. With Chapter 11, we turn to free-space optical interconnections, the most research-oriented of the topics of the book. Chapter 11, by John Neff of the University of Colorado at Boulder, discusses (in broad terms) architectural options for free-space interconnects in terms of the hardware options available. Chapter 12 by Y. C. Lee and colleagues, also of the University of Colorado at Boulder, concentrates much more on hardware implementations and the optomechanical problems existing in the area of flip-chip implementations of smart pixels.

CHAPTER TWO

Communication System Interconnection Structure

NAOAKI YAMANAKA

Large communication systems consist of several cabinets that use a bookshelf packaging structure. This chapter describes the hardware concept of communication equipment.

Figure 2.1 overviews the generic communication system. The system employs four levels of packaging structure: system level, cabinet level (bay or rack level), unit level, and board level (package level), as shown in the figure and in Table 2.1. The ICs (integrated circuits) are mounted on printed circuit boards (PCBs). About 30 PCBs are stacked into the unit. All boards are interconnected via the back plane. This interconnection is through edge connectors. A more detailed view of the cabinet is given in Figure 2.2. All electrical circuits are EMF-(electromagnetic field) shielded by metallic covers. Cabinets are electrically or optically interconnected, and the connections usually pass under the floor. Different interconnection technologies are applied in each packaging level.

For system-level interconnection, optical interconnection schemes such as SONET (synchronous optical network) or SDH (synchronous digital hierarchy) standard interfaces are used [CCI90]. The interconnection distance and speed requirements are about 400 m and from 150 Mb/s to over 2.4 Gb/s, respectively. Cabinet-level interconnection is internal system interconnection; connection distance is 5–40 m. State-of-the-art optical parallel interconnection or electrical coaxial cable interconnection can be applied here. Unit-level interconnection distances are less than 80 cm. At this level, the number of backplane interconnection links is rather large, so low cost and high form density are essential; the system clock may be as high as 80 MHz. The last level is board-level interconnection. The distances are on the order of several centimeters. For this application, noise and low power consumption are the key points.

11

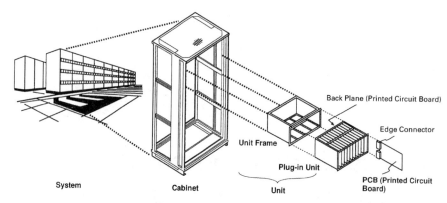

FIGURE 2.1. Communication system packaging level.

The required bit rate and interconnection distance are shown in Figure 2.3 for all four levels. Examples are given in the following sections, and detailed structures of the four interconnection levels in the communication equipment are described.

2.1 SYSTEM-TO-SYSTEM INTERCONNECTION

Between switching systems or between a switch and a transmission system, the SONET or SDH intraoffice standard interface is used. The block diagram of the SDH interface is shown in Figure 2.4. The SOH (section overhead) function includes STM (synchronous transport module) frame synchronization, error

TABLE 2.1. Communication System Packaging Level Hierarchy

Level	Contents and Example	Interconnection
I: System level	• Provide service functions (e.g., switching system)	~400 m (e.g., SDH, SONET)
II: Cabinet level	• Rack for several units	~40 m (floor)
	• Power and clock supply (e.g., processor cabinet)	~5 m (next cabinet) (e.g., coaxial cable)
III: Unit level	• Interconnect between packages	~80 cm (back plane) (e.g., 20 Mb/s V11 interface)
	• Power supply (e.g., interface unit)	
IV: Board level	• Small functions	~40 cm (On-board)
	• On-board power supply	~20 cm (between LSIs)
	• Microprocessor (e.g., 150-Mb/s SONET interface circuit)	(e.g., CMOS, ECL)

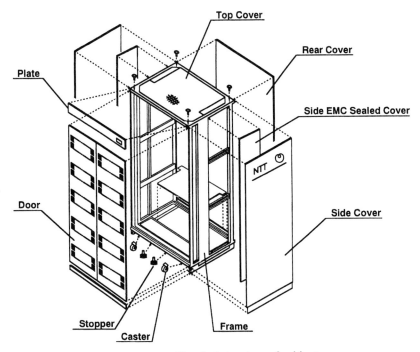

FIGURE 2.2. Detailed structure of cabinet.

detection, and switching of redundancy connections for reliability protection. This standard interface, while a little complicated, has high reliability and simplifies the connection to other systems. According to Table 2.2, all interface specifications are standardized. Figure 2.5 shows a 600-Mb/s (STM-4) module for system-to-system interconnection.

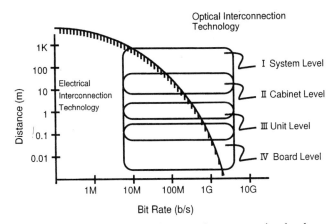

FIGURE 2.3. Communication system interconnection level map.

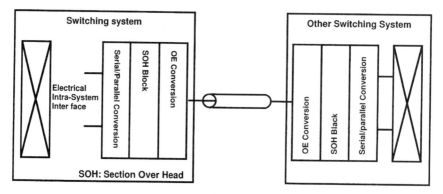

FIGURE 2.4. Block diagram of SDH interconnection.

2.2 CABINET-LEVEL INTERCONNECTION

Cabinet-level interconnections have distances ranging from 5 to 40 m. The conventional interconnection technology for this application is the 20 Mb/s×8 highway electrical interface. However, the quasi-ECL technology and the sophisticated electrical equalized amplifier technique can achieve 150-Mb/s serial-electrical interconnection [KAM93]. To achieve economical and simple high-speed interconnection for this application, multichannel optical interconnection chip sets have been reported. In this section, two state-of-the-art parallel optical interconnection technologies are described.

The first uses a 12-parallel-multimode 150-Mb/s optical interconnection chip set using an LED (light-emitting diode) array [FKI92]. Its block diagram is shown in Figure 2.6. Each interconnection module consists of an LED module and a PD (photodetector) module. The LED array is fabricated in InGaAsP. Very low current thresholds and high-speed responses of up to 600 Mb/s are achieved. On the other hand, the PD module consists of an InGaAs PIN-PD array with an 80-μm diameter. Because LED arrays and multimode fiber are used, receiving optical power is degraded to −19 dBm, but with the use of a high-sensitivity amplifier, −34.0 dBm of minimum power is guaranteed. An optical interconnection package with this chip set and specifications for this module are shown in Figure 2.7 and Table 2.3, respectively.

TABLE 2.2. SDH Bit Rate

Interface Name	Bit Rate	Channels for Telephone Connection
STM-0	51.84 Mb/s	672 ch
STM-1	155.52 Mb/s	2,016 ch
STM-4	622.08 Mb/s	8,064 ch
STM-16	2,488.32 Mb/s	32,256 ch

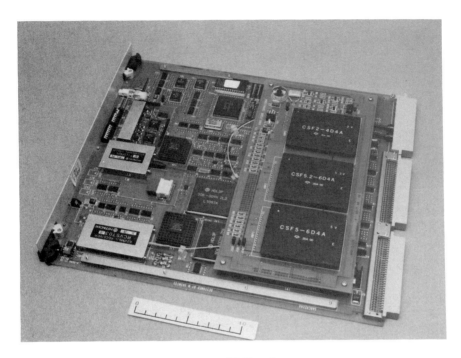

FIGURE 2.5. SDH package.

The second interconnection technology is realized as a 200-Mb/s×8 highway optical interconnection chip set using a high-speed LD (laser diode) array and a single-mode fiber array [KYT92, TAK92, DVV92]. A cross-sectional view of an eight-channel LD/PD module is shown in Figure 2.8. The module is composed of two metal blocks loaded with an SM (single-mode)-fiber array and an LD/PD

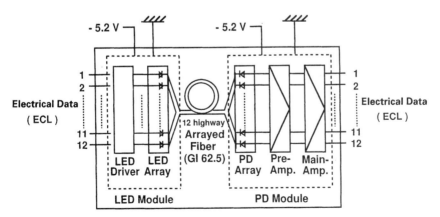

FIGURE 2.6. Schematic block diagram of LED parallel interconnection. Reprinted with permission from [FKI92].

FIGURE 2.7. Photograph of LD array interconnection module.

array, an alumina wiring plate, a metal mount, an 11-pin lead-frame metal package, a metal cap, a fiber ribbon, and a fiber connector. The module measures 30 mm(L) × 7.5 mm(W) × 5.8 mm(H).

A schematic view of the SM-fiber array is shown in Figure 2.9. Single-mode fibers are sandwiched between and positioned by V-grooved, flat silicon substrates. The SM fiber has a 10-μm-diameter core and a 125-μm-diameter cladding. V-grooved silicon substrates are used in conventional fiber array attachments. Each silicon substrate measures 6 mm(L) × 4 mm(W) ×

TABLE 2.3. Specifications of LED Array Interconnection Chip Set

		Parallel	12 HW
Sender LSI		Power supply	−5.2 V single
			3.5 W (typ.)
		Current	0–25 mA
		Speed	200–600 Mb/s
		Skew	<0.5 ns
Receiver LSI		Power supply	−5.2 V single
			1.8 W (typ.)
		Signal type	Scrambled NRZ
		Operation	0–85°C

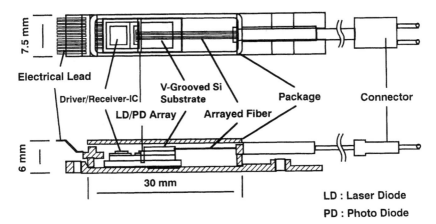

FIGURE 2.8. Structure of LD array module. Reprinted with permission from [Kat92], copyright 1992 IEEE.

0.5 mm(H). The V-grooves, formed by anisotropic etching using KOH solution, are spaced at 250-μm intervals and are 160 μm deep. By carefully controlling the orientation of the silicon crystals and the etching environment, fiber positioning accuracy can be better than 0.5 μm. Hemispherical lenses are formed on the fiber ends to improve the optical coupling efficiency between the SM-fiber array and the LD array. The lens formation process developed here is shown in Figure 2.10. The fiber ends are etched simultaneously using an HF solution and then partially melted using an electric discharger. Lens radii depend on the diameters of etched fiber ends and discharge time. Here, SM-fiber arrays with 20-μm-radius lenses are used for the LD modules. SM-fiber arrays without lenses, called *butt-end*, are used for the PD modules. In preassembly, the SM-fiber array is fused to the metal block. The block's dimensions are 6 mm(L) × 4.5 mm(W) × 1 mm(H). Good

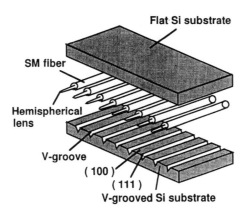

FIGURE 2.9. A schematic view of the SM-fiber array. Reprinted with permission from [Kat92], copyright 1992 IEEE.

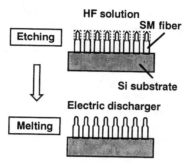

FIGURE 2.10. Hemispherical lens formation process. Reprinted with permission from [Kat92], copyright 1992 IEEE.

optical coupling efficiency between the lens-ended SM fiber and the LDs is achieved. A maximum of $-6\,dB$ optical coupling efficiency and approximately $-10\,dB$ for -2-μm misalignment tolerance are achieved.

2.3 UNIT-LEVEL INTERCONNECTION

For unit-level interconnection, the interconnection distance is less than 80 cm. The standard unit is illustrated in Figure 2.11. Printed circuit boards, on which many VLSIs are mounted, are connected to a backboard via package connectors. A total of about 900 pins are used for electrical interconnection. Due to impedance mismatching at the connector, the maximum speed of this connector is several tens of Mb/s. In other words, only several tens of Gb/s throughput is achieved by electrical interconnection.

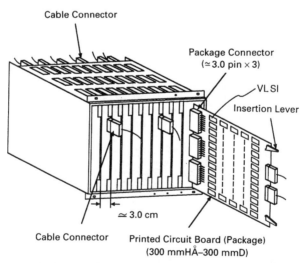

FIGURE 2.11. Unit structure of communication system.

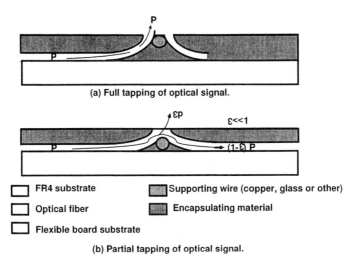

(a) Full tapping of optical signal.

(b) Partial tapping of optical signal.

FIGURE 2.12. Optical tapping technique. Reprinted with permission from [DVV92].

To achieve high-speed operation, optical backplane technologies are very important. Figure 2.12 shows two optical tapping configurations [DVV92]. Full tapping and partial tapping of optical signals are possible. Both can realize device-to-printed circuit board and board-to-board interconnection. Figure 2.13 shows one optical backplane structure that uses a planar optical connector based on the optical tapping technique. Breakthroughs for this application include polyimide optical waveguides, plastic-fiber arrays, and plastic waveguides.

2.4 BOARD-LEVEL INTERCONNECTION

For board-level interconnection, only very short distances—10 to 40 cm—are required. Accordingly, cost, size, and power dissipation are very important

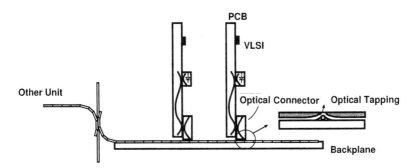

FIGURE 2.13. Schematic drawing of the interconnections in a subrack. Reprinted with permission from [DVV92].

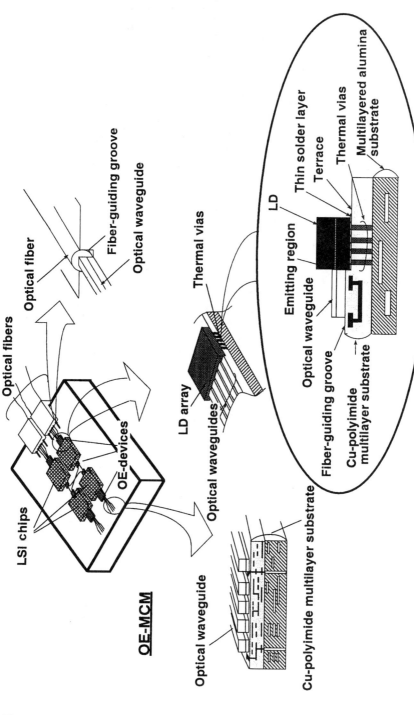

FIGURE 2.14. Schematic representation of LD packaging in OE-MCMs. Reprinted with permission from [KMT95], copyright 1995 IEEE.

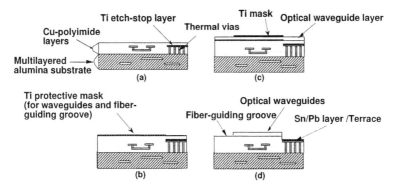

FIGURE 2.15. Fabrication processes of OE substrate. Reprinted with permission from [KMT95], copyright 1995 IEEE.

issues. Some sophisticated electrical technologies such as MCM (multichip module) and BGA (ball grid array) are used to achieve high-performance PCBs. They can achieve very complicated interconnection [YOK90, YEG95]. However, high-speed electrical interconnection also has limitations. For example, bus-type interconnection, branch, and wired-R are very difficult to realize at high speeds. Clock distribution also has big problems related to system design. Noise and crosstalk caused by electrical signals limit system throughput. These electrical limitations have been overcome with optical/electrical MCM technology [KMT95]. The state-of-the-art MCM uses copper polyimide technology. Details of its structure are described below.

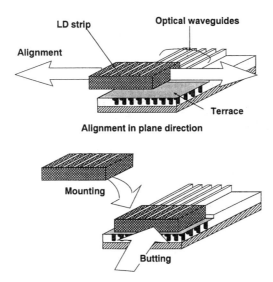

FIGURE 2.16. LD array alignment to the optical waveguides. Reprinted with permission from [KMT95], copyright 1995 IEEE.

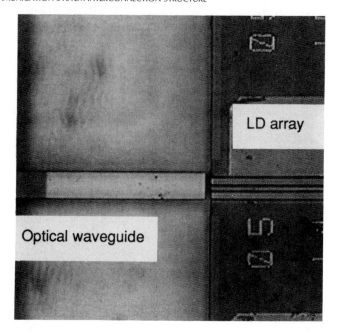

FIGURE 2.17. LD array-to-optical waveguide coupling. Reprinted with permission from [KMT95], copyright 1995 IEEE.

A schematic representation of LD packaging in the proposed OE-MCM is shown in Figure 2.14. The OE substrate is composed of fluorinated polyimide waveguides on a Cu-PI (copper-polyimide) multilayer substrate. The waveguides are 50 μm wide, with a 50-μm-thick core on a 37.5-μm-thick cladding. A junction-up LD array with an emitting height of 85 μm is bonded to the terrace and then butt-coupled to the waveguides. The surface of the precisely etched terrace is used as a standard surface for the passive positioning of the LD array in the vertical direction. The distance between the top surface of the waveguide and the terrace with a thin solder layer is set at 112.5 μm so that the LD emitting region is centered on the waveguide. The bottoms of the LD arrays are thermally connected to the alumina substrata through columnar thermal vias for heat dissipation. A 160-fold improvement in thermal conductivity under the LD bonding area is achieved by using these thermal vias. The waveguides are also coupled to the multimode fibers (Gl-50/125 μm) for intermodule interconnection by using fiber-guiding grooves [Tak93]. In order to align the fibers to the waveguides passively, the grooves are 126 μm wide and 87.5 μm deep, which is equal to half the fiber core plus the thickness of the fiber cladding. To realize these passive coupling structures, the terrace and fiber-guiding grooves should be precisely formed with different etching depths.

The fabrication processes of the OE substrate for LD array packaging are shown in Figure 2.15. In order to form terraces and fiber-guiding grooves with different depths, oxygen-resistant titanium (Ti) with a high selectivity ratio of

300 with respect to the polyimide [Shi93] is used as an etching mask or an etch-stopping layer. First, at 25 μm from the surface of the Cu-PI multilayer substrate, a Ti etch-stopping layer was deposited on the Au-metallized LD bonding area. This layer was connected to the thermal vias (Figure 2.15a). A Ti protective mask for the Cu-PI layers was formed on the surface of the Cu-PI multilayer substrate (Figure 2.15b). Fluorinated optical polyimide waveguide layers consisting of a cladding (37.5 μm thick) and a core (50 μm thick) with refractive index difference D of 1.2 were spin-coated and cured at 350°C on the Cu-PI multilayer substrate. A Ti mask for waveguide etching was then deposited on the polyimide waveguide layers (Figure 2.15c). Ridge-type optical waveguides (50 μm wide and 86.8 μm high) were fabricated simultaneously with the LD-bonding terrace (112.3 μm deep) and fiber-guiding grooves (126 μm wide and 85.4 μm deep) by reactive ion etching in an oxygen plasma. After removing the Ti etch-stopping layer by fast atomic ion beam etching, a thin Sn/Pb solder layer was coated on the terrace (Figure 2.15d). By using a highly selective Ti etching mask or etch-stopping layer, terraces and fiber-guiding grooves with different depths were simultaneously formed to within the depth error of 2 μm. Figure 2.16 illustrates the alignment step, and Figure 2.17 shows a top view of the completed package.

2.5 CONCLUSIONS

This chapter has described packaging technologies, especially optical interconnection technologies, for communication systems. The communication systems are hierarchically structured, and different package techniques are used in each level. System-to-system interconnection uses SDH or SONET optical interconnection standards. Cabinet-level interconnections use parallel optical interconnections for greater economy of space. Unit-level interconnection is achieved with an optical backplane, and optical connection between board and backplane are the key technologies here. Finally, board-level interconnection employs optical/electrical MCM technology. For this technology, chip-to-board interconnection is a very important study area.

CHAPTER THREE

Long-Distance Interconnections: The Fiber-Optic Network

GARY J. GRIMES

Optoelectronic devices for large switching and transmission telecommunications platforms for the long-haul network are packaged very differently depending on their function [Ref 91]. Devices used for interconnections within large systems are packaged differently from devices used for interconnections between large systems in a telephone switching office or between offices. Devices for intrasystem interconnection are packaged to interconnect either by means of individual optical connectors in equipment faceplates or by means of optical backplane connectors. Devices for system-to-system or long-haul interconnection are packaged for individual connections on equipment faceplates. Devices for intrasystem interconnection are usually made to be compatible with multimode glass fiber, while devices for system-to-system interconnection are usually made to be compatible with single-mode interconnection, and hence the devices are more likely to be pigtailed. Another difference is that intrasystem optical interconnection might be parallel and utilize parallel optical technologies based on optical ribbon cable, whereas system-to-system interconnection is usually serial in nature and hence uses single- or, at most, dual-fiber packaging. Active optical devices for intrasystem interconnection are typically noncooled, while thermoelectric packaging is often used for lasers for long-haul transmission. Both intrasystem interconnection and long-haul interconnection may use passive optical cross-connect systems for flexibility in configuration management, but long-haul interconnection is more likely to utilize these systems, and they tend to be stand-alone units rather than being integrated into other pieces of equipment such as transmission systems.

3.1 OVERVIEW OF OPTICAL FIBER INTERCONNECTIONS IN TELEPHONE NETWORKS

Optical fiber technology became popular for long-distance telephony applications in the 1970s. For long-distance applications, the major problems to be solved were the cost of the media and the bandwidth × distance product of the fiber media. Once the loss problem was solved, continuous processes for drawing fiber economically were quickly developed. The main problems were modal dispersion and chromatic dispersion. The modal dispersion problem was quickly solved by the use of single-mode fiber. The chromatic dispersion problem was solved by operating with narrowband sources around the zero dispersion points of the fiber. The cost of the media including installation costs over long terrestrial and undersea runs was enormous, so the cost of the terminals and the fiber terminations in particular was of essentially no consequence to the overall system cost.

The first applications for optical fiber technology in telecommunications were in intercity terrestrial links. During the early 1980s, it became apparent that optical fiber technology could economically replace some of the metallic media in the loop plant, the portion of the telephone network between central office switching equipment and the telephone sets at businesses and homes. The fiber technology available could not go all the way to the end-user sets but only to terminals distributed around neighborhoods. These terminals multiplexed the signals from a number of telephones, typically 96, and sent them on a single higher-speed digital carrier between the central office and the neighborhood terminal. The interconnections between the local switch and houses or businesses is shown in Figure 3.1, and Figure 3.2 shows how local switches are integrated into the network hierarchy of tandem and toll switches. This technology began using metallic media but switched to optical fiber in the early 1980s. The technology to terminate fibers in homes and businesses was only in the trial stage in the early 1990s, except for a few large business applications in which fiber to the business was economical. Another use of optical interconnection in telephony is in interconnection within large pieces of telecommunications switching and transmission equipment. These applications began in the early to mid-1980s in relatively low-density interconnection applications and evolved into high-density applications by the mid 1990s. The interconnections which are candidates for optical links in a local switch or PBX are shown in Figure 3.3, and the corresponding links in a transmission terminal are shown in Figure 3.4.

There are two fundamental types of facility terminations in telephone switching offices: (1) the terminations which come from telephones and other types of terminals within the local serving area and (2) the terminations which comprise the interoffice trunks that interconnect the local and long distance telephone switching offices with each other. Telephone service providers that provide local access use several distinct classes of equipment to perform these functions, but in the next decade these distinctions may blur as high-speed lightwave terminations are combined in other pieces of equipment and traditional transmission and switching functions are mixed in single pieces of equipment. One class of equipment is transmission equipment, which comprises the loop plant (the portion of the telephone network between central office switching equipment and individual telephones and terminals at homes and businesses) and terminals of the interoffice local and long-distance networks. Transmission equipment terminates several levels of the digital hierarchy. This hierarchy consists of individual lines to individual telephones, the first-level carrier systems which in North America are generally 24-channel systems and in Europe and Asia are 30-channel systems, and the higher-level systems such as DS-3 (45 Mb/s) in North America.

The highest level of functionality in the transmission plant is found in a terminal called a DACS (digital access and cross-connect system) by AT&T and a DCS (digital cross-connect system) by a number of other vendors. This piece of equipment serves to automatically reconfigure interoffice trunks or signals from the loop plant at the command of a computer terminal. The interconnection

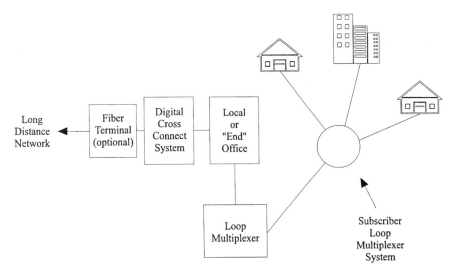

FIGURE 3.1. A local or "end" office serves as an interface between the interoffice network and the loop systems which connect the telephone system to individual homes and businesses. The telephones or other terminals in homes and businesses are attached to subscriber loop carrier terminals. The far-end loop terminal is typically located in a metal enclosure in a neighborhood. The loop multiplexer serves to multiplex signals from a number of terminals on a single wire or fiber and may take the place of the office-end subscriber loop carrier terminal. The loop multiplexer is typically connected to the local switch by means of a digital access and cross-connect terminal. This same digital access and cross-connect terminal might serve as the access path to the interoffice network, or another one might be used for this purpose. Finally, a fiber terminal, which can also typically serve as a multiplexer, terminates fibers to the interoffice network so that the local switch can send and receive calls to other offices in the local or long-distance network.

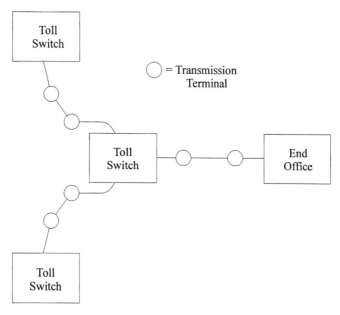

FIGURE 3.2. Long-distance offices are connected together by tandem switches. These interconnections typically consist of optical fiber facilities terminated by various pieces of transmission equipment. The tandem switches may attach to the toll or long-distance networks. If the long-distance network is handled by the same local access company—for intralata long distance, for example—then the tandem switch may handle billing or a separate toll office may be used. For the case where a separate long-distance company handles the interlata long distance, then a point of presence is used as shown.

schemes within telephone switching offices have changed dramatically over the past two decades. Before the invention by AT&T of the DACS, offices had a long free-standing wall with typically thousands or tens of thousands of metallic screw terminals to interface between the switching equipment and all the metallic facilities coming into the office. This "wall" was called a "main distribution frame." When it was decided that a facility (such as a pair of wires from a house) would terminate on another switch, the craft (telephone maintenance personnel) would determine the location of the wires and physically move them using a screwdriver. It was difficult to keep accurate records of a system of this enormous complexity with no automation, but the penalty for not doing so was a completely mysterious and unmanageable system. In modern offices using DACS or DCS equipment, a combination of optical fibers and metallic facilities terminate on the digital cross-connect equipment. Using the modern computerized switching equipment built into these systems, office moves and rearrangements can be made easily by just typing a simple instruction on the computer terminals, and physical moving of facilities is unnecessary. The cross-connects are typically of two different types. One type is used for reconfiguring the

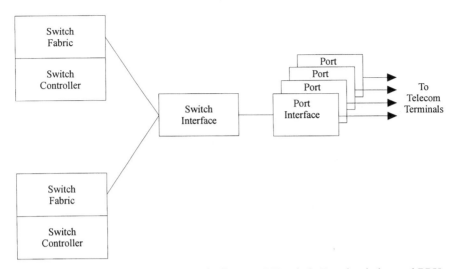

FIGURE 3.3. Duplication in a typical end office or PBX switch. Local switches and PBXs usually handle enough traffic to warrant duplication. Here a duplicated switch and switch controller are shown with duplicated bidirectional links from a switch interface. In modern equipment, these links will be optical fibers. Port interfaces serve to multiplex customer traffic from several low-speed ports or from one high-speed port and format them in such a way that the switch interface can interface them to the switch fabric.

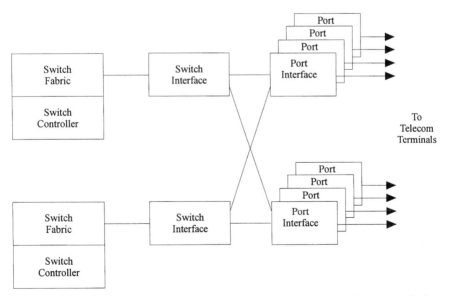

FIGURE 3.4. Duplication in a typical transmission terminal. A typical transmission terminal has a more elaborate duplication strategy. The port interfaces are cross-coupled to switch interfaces using optical fiber in the most modern systems. In a transmission, external facilities such as interoffice trunks will also be duplicated.

network around faults and does not terminate the facilities in such a way that individual subchannels are made available within the terminal. Such a cross-connect is sometimes called a "fire hose" switch because it can only switch high-speed streams of data as high-speed streams of data without terminating them, breaking them down into smaller subchannels, and rearranging individual subchannels. The other type of digital cross-connect terminal can terminate high-speed channels into individual subchannels and rearrange the individual channels onto the outgoing high-speed channels. However, these terminals often do not deal with subchannels down to the individual telephone call (DS0 or 64 kb/s) level. Another type of transmission terminal is a multiplexer, which cannot in general perform any switching function but serves only to multiplex a number of low-speed channels into one high-speed channel and also performs the reverse demultiplexing operation. Loop multiplexers are typically used in the loop plant to multiplex many individual telephone channels onto one higher-speed metallic or fiber-optic facility. The transmission which interfaces between high-speed, short–distance metallic or fiber-optic facilities within a central office and the long-distance outside plant facilities is called a *fiber terminal*. A fiber terminal can be replaced in some circumstances by long-distance optical inter-faces built into modern digital access and cross-connect systems and multi-plexers.

Switching equipment consists of local switches (sometimes called *end offices*), tandem switches which connect local switching and perhaps also terminate local loops, and toll switches which serve as the nodes and gateways for the long-distance networks. Examples of an end office or tandem switch are the AT&T 5ESS, the Northern Telecom DMS 100 and DMS 200, the GTE GTD5, and the Siemens EWSD. An example of a toll switch is the AT&T 4ESS. In addition to the call switching functions, toll and tandem switches include billing functions. These switches can switch calls at the individual telephone call level. As telecommunications systems evolve, the distinctions between these types of switches and terminals will no doubt blur and various functions may be combined into one new type of equipment, but the functions will still be present.

Optical fiber technology is also popular in distribution for cable television systems. As with telephony, optical fiber is used extensively within the longer-distance point-to-point portion of the provider network but terminates on the end user's premises set-top boxes and other terminal equipment only in advanced trials. The expensive equipment for a cable television system is located in the head end, which collects programming from satellites, combines it with locally generated programming, and drives the cable plant to the end users. Optical fibers are used to tie head ends together to avoid duplication of satellite antenna systems and to connect to local television broadcasters by means of "super-trunks." They are also used to connect head ends with neighborhood distribution systems by means of fiber "trunks." Today, most cable television systems carry analog video signals over fiber. Digital systems are emerging in order to serve interactive and multimedia needs. The requirements of analog signals place stringent return loss requirements on all fiber terminations and dramatically affect optoelectronic packaging for both active and passive components.

Cable television distribution systems are different from telephony distributions systems at present in that cable systems are typically one-way, nonswitched broadband systems, while telephone systems are typically two-way, switched narrowband systems. Somewhat ironically, most fibers in the telephone cable plant carry digital signals, while the broadband CATV fibers carry analog signals. As telephony moves into the broadband arena and as cable television moves into the two-way telecommunications arena, the services and fiber distributions are likely to become more similar.

3.2 OVERVIEW OF FIBER TERMINATIONS IN A TELECOMMUNICATIONS ENVIRONMENT

Telephone switching offices typically have several types of transmission terminals for terminating optical fiber and metallic facilities. The first and most obvious type is the termination of interoffice terrestrial or undersea fiber spans. This type of termination is typically accomplished by using discrete single-mode optical connectors on a dedicated terminal. Another type of interconnection may be a short- or medium-haul interconnection between two pieces of equipment within an office. Both of these first two types of interconnections typically conform to standards so that equipment from different vendors may be interconnected at what is called a "mid-span meet." The term *mid-span meet* is something of a misnomer because the terminations are performed typically only at the ends of the run, not at "mid-span." Single-mode fiber prevails for both of these types of interconnections, but lower-cost multimode standards for SONET/SDH may evolve for lower-cost interconnections within an office. The other types of interconnection comprise interconnection within the same piece of equipment, although pieces or modules of this piece of equipment may be separated by hundreds of meters within a building or even by several kilometers in the case of a remote switch module. These interconnections are typically optimized for cost using whatever optical technology is available, since multivendor mid-span meets are unnecessary.

All of these optical interconnections within an office give the end users as well as the manufacturers a number of important advantages compared to the previously popular metallic interconnection technologies. Interconnections of a wide range of speeds, formats, and physical facility types share the same wiring trays in an office. In North America and in most other parts of the world, these wiring trays are suspended over the equipment on hangers (metal rods) which are attached to the ceiling. Optical fibers have two major advantages over metallic interconnections. One is freedom from electromagnetic interference. Vendors of telecommunication equipment cannot possibly verify that their equipment works properly when the high-speed metallic interconnections are mixed in various configurations with a wide variety of combinations of other types of equipment from other vendors in the cable trays. Telecommunications service providers have learned that mixing multimillion-dollar (USD) pieces of equipment and sharing cable trays can lead to problems. Optical fiber takes the risk out of multivendor installations. The other big advantage of optical fiber in cable trays is size, in particular the diameter of the cable. Metallic systems typically use coaxial or shielded twisted-pair wires, both of which are more bulky than jacketed optical fibers. This is important, because even if there is initially enough room in the cable trays for wires, as moves and rearrangements occur it is generally impossible to remove a piece of wire, so each move and rearrangement means that facilities are added but not removed. The much smaller diameter of fiber makes this process much more manageable.

Another advantage of fiber is flexibility of serving arrangements. With the longer-distance capabilities of fiber, subsystems of the same piece of equipment can be spread through a floor in an office or even across different floors. Particularly in parts of Europe and Asia where land prices are high and central offices are typically cramped for space, this can save the telecommunications service provider installation costs because the present equipment does not have to be rearranged in order to make contiguous space for new systems.

3.3 PACKAGING FOR HIGH-DENSITY OPTICAL BACKPLANES

Optical backplanes refer to typically one of two levels of interconnection within large digital telecommunication switching and transmission platforms [GrH90, Nor92, Sau92]. One is a replacement for the backplane cabling, which typically interconnects racks or cabinets within a single large piece of equipment that may consist of one to fifty or more racks or cabinets. The other is a replacement for the printed wiring boards which interconnect circuit boards (now sometimes called "units") within a single rack or shelf of equipment. This first type of interconnection was introduced commercially in 1994 in the AT&T DACS VI–2000 transmission terminal with its massive internal optical interconnection. The second type of interconnection has not yet been deployed commercially but may soon be needed with the capacity of telecommunication switches and transmission systems in the tens of Gb/s range and rapidly heading higher by several orders of magnitude as platforms to support broadband as well as narrowband services are needed. Such electrical backplanes may comprise more than a square meter with 20–40 layers of interconnection. It is thought that guided-wave optics will begin to augment such electrical systems, but the timeframe is uncertain. Models for such systems have been built since the mid-1980s. Another possibility is to terminate a massive amount of bandwidth on a single circuit board, so that all switching can be accomplished on electrical buses on the board, or possibly even on the interconnects within a single integrated circuit on the board. Another possibility is to terminate all the bandwidth in a SEED device to take advantage of the enormous interconnection density possible with such free-space optics [McC94].

In order to complete the first type of optical interconnections within large telecommunication platforms, the following design issues must be addressed:

3.3.1 Optical Connector Location

For low-density applications where the telecommunications service providers must have access to the fibers for moves and rearrangements, circuit-board faceplate location of the optical connectors may be the best alternative. This has the advantage that end users have easy access to the fibers with no intermediate patch-panel arrangement, which saves optical loss and cost. However, fibers have to be routed out of the racks or cabinets by either going into a raised floor or, far more likely, into overhead trays in a central office environment. A few fibers on each board are fine so long as they are securely held in trays just above or below the shelves or subracks, but a high density of fibers could easily make it difficult to remove adjacent circuit boads without snagging and breaking fibers. Faceplate location may also have advantages for external optical connections so that only one termination is necessary to bring the fiber all the way to the optical transmitter or receiver.

Optical backplane connectors may be mixed with electrical backplane connectors to give customers the ability to plug circuit boards in and out without the need to manually remove fibers. This is a much more convenient packaging technique and leads to a much "cleaner" appearance for the equipment. It may in some cases, however, necessitate a patch panel elsewhere in the equipment to facilitate moves, growth, and rearrangement.

3.3.2 Electrostatic Discharge

With stringent new international standards for electrostatic discharge (ESD), packaging and placement of optical components, particularly optical receivers, is crucial. Central office equipment is now expected not only to survive ESD impulses but to provide error-free operation during the impulse. Accomplishing this is no easy design task with sensitive optical receivers. Often during these tests, the faceplates take the ESD impulses directly from a probe, so isolating receivers electrically from faceplates is a must. Passing the ESD requirements with optical components in the faceplates is possible. It may be advantageous if the faceplates are conductive and the portions of the connector which protrude beyond the front of the faceplate are insulated. An example of this design would be to use metallic or conductive painted plastic faceplates in combination with optical transmitters and receivers which are packaged in plastic connector housings and which have well-grounded metallic "Faraday cages" inside for the optical receivers. A less risky approach is to locate the optical components on the central part of the circuit board and to connect them to the optical connectors on the faceplate or an optical backplane connector by means of a totally nonmetallic fiber jumper consisting of a dielectric optical waveguide and its surrounding polymer protective layers. Fiber jumpers can be expensive, however, especially when they have two polished terminations. Their cost can easily exceed the cost of optical transmitters and receivers used for intrasystem optical interconnection.

Another approach to solving ESD problems is to have a good ground layer on the top layer of the board directly underneath the optical receiver.

3.3.3   Thermal Issues

Optical components generally operate better in lower-temperature environments. In order to select the best board placement for optical transmitters and receivers, complete thermal modeling of both rack-level systems and individual circuit boards using one of the commercially available computer-based graphical simulators is helpful. Optical transmitters are more efficient at lower temperatures, and optical receivers are less noisy at lower temperatures. Lower temperatures can significantly enhance reliability and product lifetimes of optical transmitters and receivers by typically a factor of two for each $10°C$ drop in temperature. Even though the thermal environment at all places on the board may lie within the range of operation of the parts chosen, good placement can enhance operation.

3.3.4   Bit Error Rate Performance Issues

Two types of optical interconnection may be present in telecommunication platforms: intrasystem interconnection and external interconnection. Even though intrasystem interconnections comprise long distances, customer expectations are that equipment will be error-free within a given piece of equipment. While external interconnections on intercity and metropolitan area facilities may be perfectly acceptable at a 10^9 bit error rate, intrasystem digital interconnections probably require a bit error rate of about 10^{15} or better to be considered "error free." Fortunately, received optical power needs to be enhanced perhaps only 2–3 dB to give the additional bit error rate performance required for intrasystem optical interconnection.

3.3.5   Maintenance of Optical Connectors

A good design guideline is to never place an optical connector where it cannot be cleaned. This is not a bad design guideline for electrical connectors, but it is especially important in optics. For faceplate-mounted optical connectors, this is not an issue. However, optical backplane connectors will generally require either rear access for cleaning, removal of the connector from the backplane to allow it to extend out near the faceplates when the circuit board is removed, or special tools to allow it to be cleaned in place from the front. The special tools may be nothing more than a long swab and the appropriate cleaning solution and a means for inspecting the finished cleaning job.

3.3.6 Selection of Optical Connectors

This job is obvious for single-fiber external connections since only a few types will be acceptable worldwide, and generally there are one or two preferred types for a given country. Examples are the ST and STII connectors of AT&T origin and the FC and SC connectors of NTT origin. For optical backplane connectors for large numbers of fibers, the choice is limited to the MACII^{TM}, but this situation may change as intrasystem optical interconnection becomes more common in commercial practice. Multifiber array connectors based on the NTT-developed MT ferrule are in widespread use in the telephone plant in Japan. The MT ferrule is a precision-molded, polymer-based material part which holds up to sixteen 125-μm-core-diameter single-mode or multimode optical fibers.

3.3.7 Single Fibers Versus Fiber Ribbons

Intrasystem optical interconnections could be much lower in cost if fiber ribbons could be used. We have found this to be impossible at this stage of development for several reasons. First, duplication in telecommunications equipment requires that fibers be fanned out to at least two different end points which are, in general, not colocated. (Otherwise, they are likely to be on the same power supplies and thus cannot achieve true duplication and high reliability.) Second, until discrete transmitters and receivers can be replaced by arrays, fiber interconnections to transmitters and receivers must consist of discrete fibers. Given the state of the art in array connector and fiber ribbon technology, this dictates that individual fibers be made into ribbons before insertion into array connectors rather than starting with ribbonized fibers and fanning out the ribbons to discrete endpoints. This greatly increases labor costs and decreases yields and is obviously one of the big problems to be solved in applying parallel optical interconnection technologies to telecommunications platforms.

3.3.8 Selection of Fiber Technologies

Selection of fiber may seem on the surface to be not too important of a choice, since often little fiber is used in a given system, and the fiber cost is often an insignificant fraction of the total cost for optical interconnection. However, the type of fiber profoundly affects the type of packaging used at every level of interconnection throughout the system, and the type of packaging profoundly affects system cost. The tolerances of both passive–passive and passive–active interfaces throughout the system are indexed to fiber diameter. Generally, optical fiber core alignment tolerances up to about 10% of the fiber diameter yield acceptable losses at each interface or connector. This means that single-mode fiber, with its 8 to 10-μm core diameter, needs an alignment tolerance of about 1 μm. Graded-index silica fiber of 50- or 62.5-μm core diameter needs an alignment tolerance of about 5–6 μm, while 1000-micron plastic fiber needs an alignment accuracy of about 10 μm. The big cost jumps in packaging occur in passive components at the tolerance limits of ordinary plastic injection molding with ordinary materials and occur in active components and active–passive interfaces (e.g., laser-to-fiber interfaces) at the threshold of going from passive to active alignment in the manufacturing process. It is obvious, therefore, that one should select the highest fiber diameter and numerical aperture consistent with the bandwidth × distance product required for the system, making allowances for future bandwidth upgradability with the existing distribution plant. With the cost of standard single-mode and multimode fibers around USD $0.50 per meter and the cost of various specialty fibers around USD $2–5 per meter, significant differences in fiber cost could occur when fiber lengths reach tens or hundreds of meters.

Two new significant choices in fibers may soon appear: graded-index plastic optical fiber (GI POF) and graded-index plastic-clad silica fiber. Koike and co-workers at Keio University demonstrated graded-index plastic optical fiber which has a bandwidth × distance product of over 2 Gb/s·km [Koi94]. While this fiber was made by drawing a preform made by an interfacial gel process, work is underway to make GI POF using a continuous process. Loss remains a problem, though, and operation is limited to short wavelengths. Deuterated fiber has demonstrated the lowest loss, but halogenated fiber is perhaps the most promising from a practical standpoint since deuterated fiber is costly. Haloge-nated polymer fiber takes advantage of replacing C–H bonds with halogen bonds having absorption wavelengths longer than the wavelengths of interest for optical interconnection.

3.3.9 Transmitter and Receiver Choices

Transmitter and receiver choices are somewhat dictated by fiber. Some impor-tant factors are cost, ESD resistance, optical connector availability, board space, mounting technique, multivendor compatibility (use of the same board footprint and signal levels), and power. Cost can be minimized by utilizing the parts near their bandwidth limit to get the most bandwidth per cost, but intrasystem interconnection generally requires error-free performance. This means that the overall power budget needs to be derated to give a bit error rate of 10^{15} instead of the more usual 10^9. The system power may be optimized by using discrete devices, since completely packaged transmitters and receivers generally do not allow system designers to choose an optimal power-performance operating point.

3.3.10 Wavelength-Division Multiplexing

Wavelength-division multiplexing is usually done when media (fiber) costs are large compared to terminal costs. This naturally occurs in long-distance applica-tions, in which terminal costs are not as important as utilizing expensive long fibers run to the maximum extent possible. Since fiber costs are typically low in optical backplane applications, WDM techniques are generally out of the question, since the cost of a single WDM device may dwarf the cost of other system components. WDM devices may be economical in those cases in which the backplane area for optical interconnection must be kept to a minimum in order to allow for massive electrical routing on the backplane. Putting multiple signals on each fiber reduces the number of fiber positions needed in a connector and the overall connector size.

3.3.11 Timing Distribution and Recovery

For short-distance parallel interconnection, clock distribution rather than clock recovery may be advantageous. Minimizing skew between fibers in fiber ribbon cables is an important research topic, and results will be of great interest to systems designers of both telecommunications and supercomputer platforms. For those who must recover clock, Manchester encoding is a good choice since bandwidth is cheap and available. For those who must utilize bandwidth to avoid going to much higher transmitter and receiver costs, SAW devices and phase-locked loops are reasonable choices. SAW devices are relatively trouble-free, while custom designed phase-locked loops may require extensive testing and lengthy design iterations.

3.3.12 Modal Noise and Reflections

Modal noise, which is noise caused by time-varying interference in coherent or partially coherent beams, and reflections can cause problems. Modal noise is only a problem in multimode optical systems, but reflections can cause problems in both single-mode and multimode systems. Reflections have only recently begun to be of concern in short-distance optical communications because of the more common use of lasers and because of the higher speeds involved. Both problems can be minimized by cutting down on the number of interconnections in systems by the use of physical contact (PC) connectors with no dielectric–air interfaces.

Modal noise can be aggravated by connectors and by multimode optical couplers (splitters and star couplers) which work on the principle of dividing mode volumes. Unless special care is taken to make couplers non-mode-sensitive, the mode partitioning can cause noise. Effective ways to avoid modal noise are to use power budgets well in excess of the power required to give the desired bit error rate and to use waveguides with many rather than a few modes so that any anomalies in any given mode or group of modes will not be significant to the power transmitted on any given optical path. The first solution can add cost and power to the system, and the last solution can decrease the bandwidth × distance product of the system, but product design cycle times may be critical to hit a market window and therefore may be more important to the overall success of a design.

3.4   AN EXAMPLE OPTICAL BACKPLANE

The first optical backplane deployed in a large-scale telecommunications platform or supercomputer was deployed in 1994 by AT&T in the DACS VI-2000 digital access and cross-connect system designed for the SDH synchronous transmission standard [Gri93, GPL92]. The design was based on the MACII™ 18-position optical backplane connector. This connector allowed a great deal of bandwidth to be extended through the backplane with a minimum penalty of routing space in the large electrical backplane. In a small area, 12 of the 18 MACII fiber positions were used to carry 155 Mb/s each through the backplane on standard 62.5-μm core-diameter graded-index silica multimode fiber.

The functionality of the board is shown in Figure 3.5. The optics side of the board is shown in Figure 3.6, and the electronics side of the board is shown in Figure 3.7. This board terminates four bidirectional 155-Mb/s SDH streams in a manner which precisely frame aligns the streams in time on an electrical backplane. The optical components shown in Figure 3.6 can be identified easily by using Figure 3.8, which shows the principal features of the layout. The multifiber array connector, the four 1 × 2 fused biconical tapered splitters, the four clock recovery chips identified by "CR," the four optical receivers identified by "R," the four optical transmitters identified by "T," and the bundles of discrete fibers are shown in the same positions as they appear in the photograph of the circuit board. The fiber routing is done by holding the discrete fibers in small clips. The clips hold the fibers in place and prevent them from snagging components on adjacent circuit boards and from getting caught in the card guides during circuit

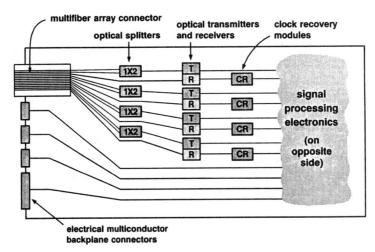

FIGURE 3.5. Partitioning of components on the example optoelectronic circuit board. The 2 × 2 couplers are packaged on a panel off the board. The signal processing electronics are shown in Figure 3.7. Reprinted with permission from *Proceedings of the 43rd Electronic Components and Technology Conference* ([Gri93]), copyright 1993 IEEE.

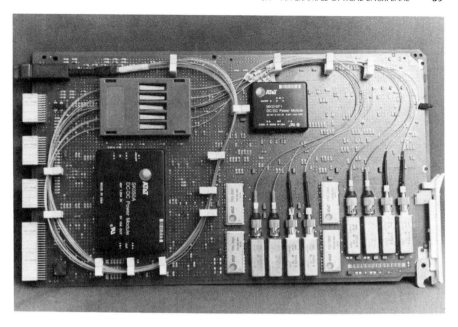

FIGURE 3.6. The example optoelectronic circuit board, optics side. The MACII™connector is shown with its dust cover at the upper left. The four white board-edge connectors under it are the electrical backplane connectors. The optical components are easily identifiable using Figure 3.8. The two large black rectangular objects are DC/DC power converters. A large number of surface-mount passive electronic components are also shown. The larger electronic components are on the opposite side of the board shown in Figure 3.7. The fiber bundles are held in place by the 13 white plastic clips. Reprinted with permission from [Gri93], copyright 1993 IEEE.

board insertion and removal. All fibers were 62.5-μm-core-diameter multimode fibers with 900-μm jackets applied in a loose tube-type construction. The loose tube construction was used for two reasons. First, the jackets were put in place on the optical couplers after the fusing process. Second, it would have been impossible to strip the ends of the fiber several centimeters, which is required for placement in the MACII connector if a tightly bound jacket had been used. The fibers were formed into ribbon cables by a manual process, and the resulting ribbon cables were inserted into the MACII.

The MACII Quick Release connector is shown in Figure 3.9. This is the backplane portion of the connector, which easily snaps out of the backplane for cleaning. The circuit board or unit half of the MACII connector is shown in Figure 3.10.

This first design of a high-density optoelectronic circuit board for production was effective in the system application using discrete transmitters and receivers,

FIGURE 3.7. Optoelectronic circuit board, electronics side. The largest components are surface-mount CMOS semicustom ICs. The large blank area in the lower left is reserved for mounting the optoelectronic components on the opposite side of the board. The multifiber array connector protrudes slightly past the upper right side of the board, and the board latch is at the lower left. Reprinted with permission from [Gri93], copyright 1993 IEEE.

discrete optical fibers, and discrete optical splitters. The key to the success of this design was the flexible serving arrangements and long-distance remoting of pieces of the finished system and the good use made of the most precious system resource, the printed backplane. The optical backplane which resulted allowed a product very different from those of competing systems. While the competing systems had bulky metallic cable harnesses which constrained the systems to be deployed in contiguous configurations, the DACS VI-2000 digital access and cross-connect system could be deployed in an office on a shelf-by-shelf basis, except for portions of the core switching and control hardware, because of the optical interconnections used between the port interfaces and the core switching portions of the system.

3.5   PACKAGING EMERGING TECHNOLOGIES FOR PARALLEL OPTICS

Optical backplanes can greatly benefit from array-style optics. The first array-style backplane connectors, the MACIIs, could be terminated with fiber ribbon

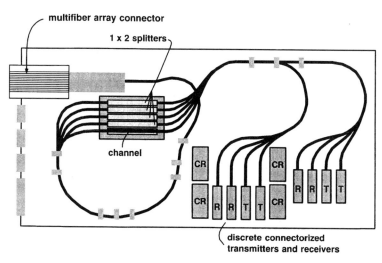

FIGURE 3.8. Simplified drawing of the example optoelectronic circuit board, optics side. The optical transmitters are labeled T, the optical receivers are labeled R, and the SAW clock recovery units are labeled CR. The multifiber array connector is shown in a larger scale for visibility. The molded plastic coupler holder is shown with its snap-on lid removed. Reprinted with permission from [Gri93], copyright 1993 IEEE.

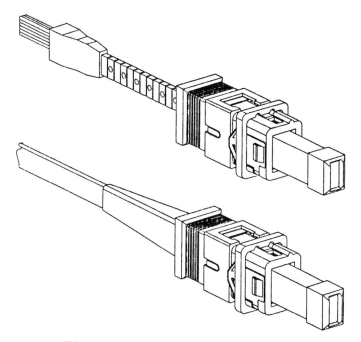

FIGURE 3.9. MACII™ Quick Release backplane connectors. The top version terminates an optical fanout cable, and the bottom version terminates an optical ribbon cable. Reprinted with permission from [Gri93], copyright 1993 IEEE.

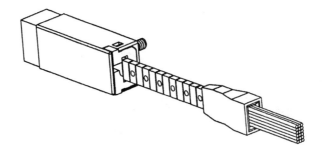

FIGURE 3.10. MACII™ circuit-board connector with individual fiber fanout assembly. Reprinted with permission from [Gri93], copyright 1993 IEEE.

cable, but there were no array-style sources and detectors with which they could be mated. This meant that discrete transmitters and receivers had to be used; since fiber ribbons cannot in general be fanned out into discrete fibers, ribbons had to be made from individual fibers. This is a costly manual process. With the emergence of both array-style sources and detectors, ribbon cables can be used on circuit boards, eliminating discrete fibers. Following the example circuit board discussed in the previous section, the packaging might appear as shown in Figure 3.1. Figure 3.11 shows the array-style optical backplane connector to still allow blind mating of the circuit boards, with source and array detectors attached by means of optical fiber ribbons. The fiber ribbon assembly will be a

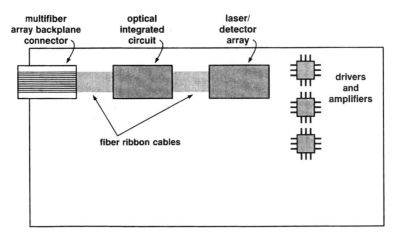

FIGURE 3.11. In this embodiment of an optoelectronic circuit board, the multifiber array connector is attached to source and detector arrays by means of custom fiber ribbon cable assemblies. The fiber ribbon cable assemblies could be replaced by the use of Optiflex polymer carriers of optical fibers. Reprinted with permission from [Gri93], copyright 1993 IEEE.

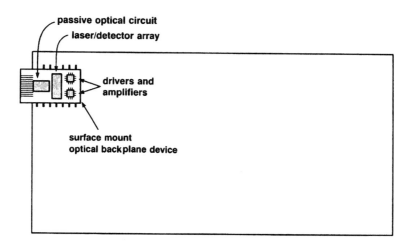

FIGURE 3.12. Eventually, high-density optical interfaces will interface directly with logic level signals on the electrical side and directly with a multifiber array connector on the optical side, all in very little circuit-board space. The source and detector arrays and the interface electronics will all be packaged inside. It will probably be surface-mount for the electrical contacts but require some other stronger mechanical means of attachment for the board so that mating the backplane optical connector will not damage the electrical contacts. Reprinted with permission from [Gri93], copyright 1993 IEEE.

custom assembly in many cases. An alternative for the fiber ribbons are discrete fibers embedded in a two-layer "sandwich" of polymer sheets held together with an adhesive. This assembly is made by AT&T and is called Optiflex^{TM} [HBS93]. Optiflex provides for easy attachment of multifiber connectors to groups of fibers.

The next logical evolution is shown in Figure 3.12. Here a single surface-mountable array-style optical backplane connector contains both the source and detector arrays along with the associated electronics. This gives systems designers the ability to utilize high-density parallel optics without being concerned with many of the packaging problems associated with today's designs and with very little required board space. This approach could almost double the amount of VLSI and functionality placed on a circuit board. This in turn could greatly reduce the bulk and footprint of the finished system, saving the customer valuable space in the central office or computer room, depending upon the application. The technologies for integrating active devices with an optical backplane connector include silicon optical bench technology [Nor93] and polymer waveguide technology [Boo89].

CHAPTER FOUR

What a Packaging Engineer Needs to Know About Optics

FELIX KAPRON and ROBERT HOLLAND

A possible alternative title to our chapter here could be *coupling light between alternative media,* which to a lighting engineer might be termed "light piping" or "applied radiometry." The point is that the salient feature of today's systems is really the power throughputs. As system sophistication increases, this may change, but after 25 years of low-loss fibers, coupling efficiency remains the prime concern and thus will be the concern of this chapter on what the optical engineer needs to know.

4.1 CHARACTERISTICS OF GOOD COUPLING

In our context, an optical coupling system (and the majority of systems we know of are such) accepts optical power from one optical component, modifies it somehow, and transfers the power to another optical component. In this power transfer there are two factors that stand out. One factor is the efficiency with which the power is transferred. Hence, if P_1 is the power into the coupling system from one component and P_2 is the power into another component, then one defines the *transmittance* as

$$T = \frac{P_2}{P_1}.$$

This is a number less than unity and may be expressed as a percentage. Physically, this is not unity because of (1) absorption, the conversion of photons into phonons and possibly longer-wavelength photons, and (2) scattering, the conversion of "guided" light into "unguided" light. In a cascade of coupling systems, clearly the overall transmittance is the product of the individual transmittances.

In several areas of application, one defines the *insertion loss* as

$$IL = -\log_{10} T \quad \text{in (positive) dB.}$$

Convenient numbers to remember are that transmittances of 100%, 80%, 65%, 50%, and 10% correspond to insertion losses of 0 dB, 1 dB, 2 dB, 3 dB, and 10 dB, respectively. From these, one can construct other common combinations. A useful function of this formulation is that in the cascade the overall insertion loss is the sum of the individual insertion losses. Particularly in fiber-optic systems, losses may be "discrete" at interfaces or components or "continuous" along optical fiber.

Another factor is the efficiency with which power is reflected back into the first component. Hence, if P_3 is the power so reflected, classical optics defines the *reflectance* as

$$R = \frac{P_3}{P_1}.$$

Again, this is a number less than unity and may be expressed as a percentage. Physically, this reflection is due to refractive index mismatches between the components and coupler. The same cascading rule holds if the path of the reflected light is traced through the optical system. Also, one defines the *(optical) return loss* as

$$RL = -\log_{10} R \quad \text{in (positive) dB.}$$

As with losses, reflections may be discrete or continuous.

In fiber optics, reflectance is sometimes the negative logarithmic form taken effectively to apply to a single discrete optical component. Optical return loss is then applied to a cascade of discrete and continuous reflections.

The disadvantage of high insertion loss in an optical system is obvious, and it can be counteracted by a better optical coupling process, by a more powerful optical source, by a more sensitive optical detector, or by optical amplification in between. The problems with and solutions for reflections are not as straightforward. Single and multiple reflections may occur in fiber-optic systems to cause noise at the transmitter and spurious signals at the receiver, possibly made worse by optical amplifiers. These are minimized with the use of index-matching materials, optical coatings, angled interfaces (to deflect the light from the transmission path), and optical isolators. The latter device rotates the polarization of light passing through it so that impinging reflected light is further rotated and blocked.

In fiber optics, light reflected from splices or connectors or at transmitter or receiver interfaces can limit the rate of information transmission due to inherent noise caused by excess light in the system. For digital systems, reflectances of

$-40\,dB$ are readily available with most components and are sufficient for practical bit error rates. For amplitude-modulated analog systems, values of $-55\,dB$ are desirable.

4.2 SPOT SIZE OR MODE-FIELD DIAMETER IN WAVEGUIDING STRUCTURES

A term often used in coupling calculations is *spot size* (SS), given the symbol w_0, that originated with the propagation of cylindrical gas-laser beams. In fibers, these have a cross-sectional field intensity profile that is circularly symmetric and close to Gaussian along a radial coordinate r. Sometimes more useful is the radial power density profile given by

$$p(r) = p(0)\exp\left\{-2\left(\frac{r}{w_0}\right)^2\right\},$$

where now $2w_0$ is called the *mode-field diameter* (MFD). Clearly, the mode-field *radius* (MFR) equals the spot size, which is in turn the root-mean-squared (RMS) width of the distribution. The term MFD originated with single-mode optical fiber. As shown in Figure 4.1, an ideal *step-index* optical fiber consists of a

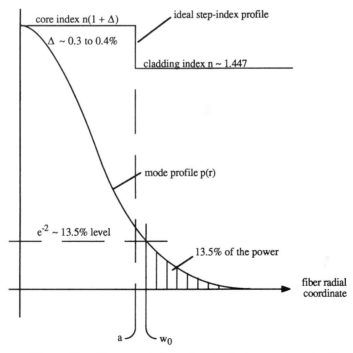

FIGURE 4.1. Waveguide with a step-index refractive index profile and with a Gaussian mode profile. The relationship of core and mode-field radii is shown.

core of refractive index $n(1 + \Delta)$, typically made of a doped silica. It is surrounded by a *cladding* of lower refractive index n, typically made of pure silica and having an outside diameter of $125\,\mu$m. The relative refractive index difference Δ is about $1/3\%$, and the core diameter (CD) $2a$ is about $8.5\,\mu$m.

There are fibers of different refractive index profile design, fibers that are multimode rather than single-mode, and fibers made from soft glasses or from plastics. We will concentrate on the most common silica-based step-index telecommunications fiber to illustrate some properties of the MFD. These properties are found also in other waveguiding structures such as semiconductor lasers and planar waveguiding structures, both passive and active. The MFD plays a key role in coupling among the components. Some power profiles deviate from Gaussian, and others become elliptical, but the general principles still hold.

One principle as shown in Figure 4.1 is that, at the circle formed by the mode-field radius, the power intensity is e^{-2} or 13.5% of the intensity at the center. Also, the same significant percentage of power is contained by the Gaussian profile outside this ring.

Another principle is that the MFD is larger than the CD. In fact, the ratio of the two can be expressed in the form

$$\frac{\text{MFD}}{\text{CD}} = A + B\left(\frac{\lambda}{\lambda_c}\right)^D,$$

where

$$\lambda_c = Ean\sqrt{\Delta}$$

is the theoretical *cutoff wavelength* of the waveguide and A, B, D, and E are empirical fit parameters. The operating wavelength is λ, and when it lies below the cutoff wavelength, the fiber or other waveguide is no longer single-mode. (In practice, the operating cutoff wavelength may be lower since loosely bound higher-order modes near theoretical cutoff are easily scattered away by bends and stripped away by the plastic jacket surrounding the glass fiber.)

The values of the parameters are not important in our context, but they predict that the MFD is about 6% larger than the CD, or about $9\,\mu$m, at an operating wavelength of 1310 nm. Moreover, the mode power profile gets wider with increasing wavelength, so that at 1550 nm the MFD is about 20% larger than the CD, or about $10.25\,\mu$m. Hence, the MFD grows about 14% between 1310 nm and 1550 nm, compared to the wavelength increase of 18.3%.

Another principle is that significant power lies outside the guiding core. The fraction of total power outside the core is

$$F = \exp\left\{-2\left(\frac{a}{w_0}\right)^2\right\},$$

which is about 17% at 1310 nm and 25% at 1550 nm. This indicates that at longer wavelengths the optical power is more loosely bound to the structure and is more susceptible to bending and being radiated away. This is one limit as to how small waveguiding components can be made.

As the mode exits the end face of an optical fiber, planar waveguide, or laser diode, the mode-field radius or spot size expands with distance z as

$$w_0^2(z) = \left[w_0^2 + \left(\frac{z\lambda}{\pi w_0} \right)^2 \right].$$

In the far field, the mode profile in the spatial-domain Hänkel transforms into a profile in the angular domain. Gaussian profiles give

$$P(q) = P(0) \exp\left\{ -2\left(\frac{q}{W_0} \right)^2 \right\},$$

where

$$q = \frac{2\pi}{\lambda} \sin\theta \quad \text{and} \quad w_0 W_0 = 2.$$

From this last relation, the *angular rms width* in degrees is

$$\theta_0(°) \cong 18 \frac{\lambda}{w_0}.$$

Because of the growth of MFD between 1310 nm and 1550 nm, the angular width increases only about 1.3% between the two wavelengths. Again, it should be emphasized that, in addition to wavelength variation, MFD will be different for different single-mode waveguiding structures. Fiber examples are dispersion-shifted fiber ($\cong 8.25\,\mu\text{m}$ at 1550 nm), nonzero dispersion fiber, dispersion-compensating fiber (both $\cong 5\,\mu\text{m}$ at 1500 nm), and active optically amplifying fiber (less than $4\,\mu\text{m}$ at 1550 nm). These mode profiles are nominally circular; elliptical profiles are characteristic of polarization-preserving fiber and sometimes planar waveguides and of laser diodes which are also smaller (e.g., $1.5\,\mu\text{m} \times 3.5\,\mu\text{m}$). Moreover, the profiles may deviate from a Gaussian shape. We will not consider multimode coupling in this chapter, but it is interesting to note that, for multimode fiber, CDs may typically range from $50\,\mu\text{m}$ to over $200\,\mu\text{m}$. The far-field angle is defined by the *numerical aperture* (NA) or sine of the half-angle of the exiting light. This angle can vary with source excitation conditions at the beginning of the fiber and with the fiber length, quite unlike the behavior of single-mode fiber. Values may range from 0.15 to over 0.3. Also unlike the situation for single-mode fibers, the far-field angle is independent of the CD and of wavelength (to first order), and MFDs are not defined.

4.3   COUPLING LOSSES DUE TO MFD MISMATCHES AND OFFSETS

For efficient transfer of optical energy from one optical component to another, their respective mode profiles should "overlap" as much as possible. This criterion leads to a general formula [And94] that relates MFD size mismatches as well as offsets that are transverse (with respect to the optical axis), long-itudinal, or angular. Of practical use are several simple approximations.

First, consider size mismatch in single-mode components [Lad93]. Define the ratio R of the difference of MFRs (SSs) w_1 and w_2 over their average:

$$R = \frac{2(w_1 - w_2)}{w_1 + w_2}.$$

Then the dB loss at the interface between the mode fields is, to a good approximation,

$$L(R) \cong 4.343R^2 \quad \text{in dB.}$$

Next, consider a transverse offset x. Then

$$L(x) \cong 4.343\left(\frac{x}{w_0}\right)^2 \quad \text{in dB.}$$

For a longitudinal separation z, the loss is

$$L(z) \cong 5.3\left(\frac{\lambda z}{10w_0^2}\right)^2 \quad \text{in dB.}$$

Hence, for an MFD of $9\,\mu$m at 1310 nm, a 0.1 dB loss corresponds to a separation of 21 μm. In comparison with the previous example, it is obvious that coupling is much less sensitive to longitudinal separation than to transverse offset. Finally, for an angular tilt of $0°$, the loss is

$$L(\theta) = 2.7\left(\frac{w_0\theta}{10\lambda}\right)^2 \quad \text{in dB.}$$

Table 4.1 gives numerical values for these "coupling errors" when the tolerable loss to any one error is 0.1 dB. For such small errors the losses are additive, and for large errors, they are interactive, and the full formula [And94] must then be used. Note that loss is particularly sensitive to transverse and angular offsets. Moreover, these losses are bidirectionally equal, a fact that may counter intuition in picturing the coupling from a large MFD to a small MFD and vice versa. Another way of picturing it is to think of the counterbalancing effect of a small angular distribution into a large one.

With multimode fiber and other multimode devices, the coupling from a

TABLE 4.1. Coupling Errors Corresponding to 0.1-dB Loss

Coupling Error Type	Error Magnitude
MFD mismatch	± 15%
Transverse offset	7.5% of MFD
Longitudinal offset[a]	21 μm
Angular offset[a]	0.56°

[a] Assumes a mode-field diameter of 9 μm at wavelength of 1310 nm.

Lambertian (isotropically bright) source increases as the term $(CD \times NA)^2$. In practice, more collimated sources reduce this dependency. Moreover, losses are now directional. In the spatial domain, a small-core fiber will "perfectly" couple into a large-core one, whereas in the reverse direction the loss is related to the transmittance ratio (small CD/large CD)^2. Analogous statements hold for the NAs in the angular domain.

Launching power into multimode devices can be lossy if leaky higher-order modes are excited. With multimode fibers, a length-dependent transient loss is experienced with "overfilled" light launching, and the same is true on a smaller scale with multimode components. Moreover, with a very coherent source, modal noise in the coupling to the fiber or component can degrade the signal to be transmitted. Single-mode-to-multimode coupling is very efficient, whereas the reverse direction presents challenging problems [CYY95]. This is the case whenever one tries to couple from a larger to a smaller mode volume. The former is sometimes used for efficient coupling from a single-mode fiber to a multimode pigtail to a detector.

4.4 LASER DIODE-TO-FIBER COUPLING

The brute-force approach of simply butting an optical fiber almost against the laser diode allows for only about a 10–20% coupling efficiency, or a 7- to 10-dB coupling loss. Part of this is due to the fact that the emission pattern is somewhat elliptical, as mentioned earlier. Analytical expressions for such patterns have been used for both light-emitting diodes (LEDs) [CRS87] and laser diodes even recently [Lad93,Nem94]. Figure 4.2 schematically shows some approaches to raising the efficiency and reducing the loss. Some of these schemes use lenses that may be cylindrical rather than spherical to compensate for this. One consideration in these arrangements is minimizing the package size. Another is optimizing the alignment. Figure 4.3 shows the sensitivity of several schemes to lateral displacement between the diode and the lenses or fiber. Longitudinal displacement is another factor. Tuning the parameters reveals that, generally, the lower the coupling loss, the more sensitive the optimum coupling is to displacement.

Another consideration is reflection of the emitted light back into the device.

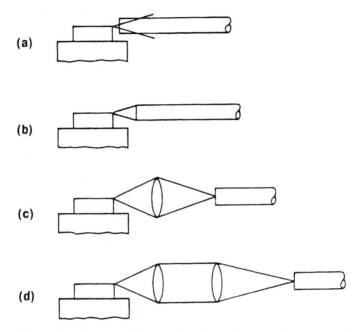

FIGURE 4.2. Comparison of four techniques of coupling a laser diode to a fiber: (a) cleaved fiber coupling; (b) coupling with a fiber lenslet; (c) using a single-lens imaging system; and (d) using a double-lens imaging system. Copyright 1989 AT&T, from [AIB89].

Planar surfaces perpendicular to the optical axis are particularly bad in this respect. This can be countered by having such surfaces at an angle from the perpendicular; because of the small angular spread of a single-mode beam, this angle is typically less than 10°. Another method is to have curved surfaces. As Figure 4.3 shows, the radius of curvature must be small because of the small MFD of single-mode fiber or components. Finally, an index-matching material between the surfaces or thin-film coatings have been used.

Still another related consideration is the mechanical stability of the coupling arrangement, especially to vibration or temperature fluctuations. For this reason and others, the most common means of coupling laser diode light into a single-mode fiber is via a lensed and often tapered fiber. Such tapers adiabatically change the MFD of the fiber. They may be fabricated by pulling on a short length of fiber; this reduces the diameters of both the core and cladding, as shown in Figure 4.4a. By a formula in Section 4.3, however, this enlarges the MFD. If a stub of the fiber is etched, part of the cladding is removed, so that the core diameter and MFD remain about the same as in Figure 4.4b. In either case, a lens at the taper tip is desired to reduce the MFD for better coupling. Losses are typically 3–4 dB. Such lenses may be formed on the fiber end by either (a) photolithography, (b) etching, (c) heating by a flame or electric arc or laser, (d) dipping into molten glass and fusing, or (e) wedge-polishing.

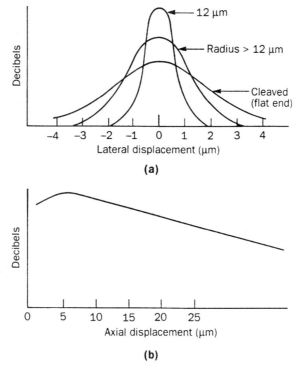

FIGURE 4.3. Plots illustrating the sensitivity of laser diode-fiber coupling efficiency to (a) lateral displacement and (b) axial displacement. Copyright 1989 AT&T, from [AIB89]. All rights reserved.

The elliptical nature of the diode laser beam pattern necessitates departures from a simple hemispherical surface. When a hyperbolic lens [EPD93] was formed on a fiber end with a pulsed CO_2 laser beam, the transverse offset loss was about the same as in the hemispheric case, but there was less sensitivity to tilt and more sensitivity to longitudinal translation. With more complex shapes, these

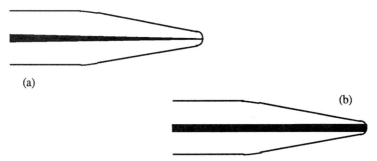

FIGURE 4.4. Tapered lensed fiber tip with (a) core and cladding tapered and (b) only cladding tapered.

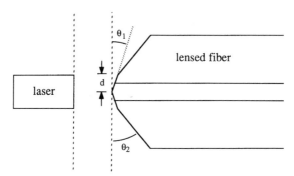

FIGURE 4.5. Lensing by a double-wedge fiber end. Modified with permission from [MoW95], copyright 1995 IEEE.

sensitivities can be expected to also be more complicated. Nevertheless, a coupling efficiency of 90% or 0.45 dB loss was achieved. Reflectance was −40 dB, or −60 dB with a coating—only about 4 dB worse than with a hemispherical lens. Another approach is a wedged-shaped fiber end [Sha90], for which a recent improvement [MoW95] has been a double-wedge lens. It is formed by grinding the tip of an untapered fiber, and the shape in one plane is shown in Figure 4.5. Perpendicular to this plane, the fiber and laser modes are closely matched. The theoretical coupling prediction was 96% or 0.18 dB, compared to a measurement as good as 80% or 1 dB.

4.5   FIBER-TO-PLANAR WAVEGUIDE COUPLING

Coupling to the guided optical modes of a waveguide structure is necessary in experiments to measure fundamental properties as well as in the practical operation of real devices. Channel waveguiding devices first appeared primarily to provide passive functionality, and there has recently been growth in their use for active applications. Components such as couplers, filters, multiplexers, splitters, and taps have all been integrated into optical waveguide formats, typically through the process of multilayer film deposition on a substrate material. For single-mode guides, all four of the coupling loss mechanisms discussed in Section 4.3 will play a role in fiber-to-waveguide packaging. The highest coupling efficiency will occur when the launching field distribution closely matches that of the receiving field. Due to the greater freedom in adjusting passive waveguide parameters during design and fabrication (refractive index, thickness, channel width, etc.), optimization is somewhat less difficult than with laser-to-fiber coupling where active device attributes must be simultaneously met in the design.

Typically, one chooses to design the waveguide so that its mode field is close to the fiber fundamental mode. If this is done, butt coupling of the fiber end to the waveguide is the preferred method because it largely eliminates one loss mechanism, that of longitudinal mismatch. Since the modes of both the fiber and waveguide are planar in nature, butt jointing ensures that the shapes of the wavefronts are matched at the endfaces. Otherwise, if the light emerging from the fiber was allowed to diverge, a lens or other focusing instrument would be required to reshape the wavefront. Thus, for butt jointing, spot-size misalignment and geometrical mismatch (axis and tilt) are the dominant challenges to fiber-to-waveguide coupling.

Many waveguide devices today are fabricated of doped silica deposited on a substrate. Through this process, the refractive index of the guiding core and cladding layers can be adjusted so that the waveguide spot size mimics that of the fiber's fundamental mode. However, this can call for the deposition of very thick film layers (50 μm or greater) and creates some fabrication challenges. Fortunately, the exact shape of the waveguide channel has little effect on its coupling to a single-mode fiber. If a square channel is created, the rectangular modes of the waveguide are very similar to the circular modes of the fiber, and the Gaussian beam equations presented in Section 4.3 can be applied. Currently, waveguide devices are typical in glass-on-silicon, glass-on-glass, and all polymer systems.

For some active devices such as switches and modulators, the material system used in the active region also forms the basis for waveguiding input and output channels. Unfortunately, most materials which exhibit desirable electrooptic behavior (Kerr effect, electroabsorption, etc.) often make it difficult to tailor the waveguide mode properties exactly to those of a fiber. Materials such as semiconductors and engineered polymers typically exhibit refractive indices much higher than those of glass, producing single-mode waveguides with much smaller geometrical dimensions and much larger Fresnel reflection losses. The beam-shape issues discussed in Section 4.4, minor for passive guides, become significant for active guides. Therefore, some means of adjusting the spot size such as the use of lensed fibers is usually adopted for such waveguides.

One exception to this is in cases where a passive waveguide can function as the "lead-in" to an active device. For an optical source device, waveguide channels would be required to transport light away from a laser. For an optical receiver, channels would be used to carry light onto a detector. In these cases, waveguide-to-device coupling is built into the structure and the external fiber must only be coupled to the "lead-in" guide. Especially for receiver applications where the detector active region is on the order of 50–100 μm in diameter, multimode waveguides are attractive because they relax the coupling requirements. Although geometrical mismatches are reduced, issues such as angular offset and numerical aperture (NA) are still of concern since one wishes to avoid noise created by mode interference when higher orders are excited selectively.

With any waveguide, coupling losses depend primarily on the integrity (strength and alignment accuracy) of the attached fiber. A common approach is to use mechanical registration of the fiber end for coarse alignment and support. While this registration (as part of the waveguide structure) can provide final alignment accuracy for multimode applications, single-mode devices typically require additional active alignment of the fiber to the waveguide. A V-groove or notch fashioned into a mounting fixture is one popular approach for providing this mechanical registration. If sufficient accuracy can be fabricated into the fiber-to-fiber pitch of the registration fixture, passive coarse alignment for several fibers at a time is possible. Unfortunately, many waveguide devices are passive in nature, and active alignment still requires the injection of light through the system and fine control of final alignment. Once aligned, the fiber must be fixed or attached to the waveguide structure. Common approaches are to use epoxy adhesives, weld bonding, or some other means. Drawbacks to the use of epoxy are that final alignment must be maintained during cure. Also of concern is the subsequent relaxation or moisture swelling of the epoxy serving to displace the fiber-to-waveguide alignment. Also of concern with metal bonding or welding is movement created during the fixing process.

4.6   FIBER-TO-FIBER COUPLING

In applications such as splicing or connection where coupling is required between two similar fibers, two loss mechanisms (wavefront and spot-size mismatch) are greatly minimized in all cases. The challenge of fiber-to-fiber coupling thus reduces primarily to that of a butt joint between two flat faces and any geometrical mismatches that may occur. The formulas of Section 4.3 may again be applied, and naturally for two fibers with similar spot sizes, the coupling loss is small. For two fibers with different spot sizes, some method of beam transformation (lens, taper, intermediate fiber length, or graded medium) can be used to adjust the spot size. Longitudinal offset remains small as long as the two fibers can be brought together within a Rayleigh length. Therefore, fiber-to-fiber coupling losses are largely based on lateral mismatches and reflections. Geometrical offsets (transverse and tilt) dominate in fiber-to-fiber coupling losses. It should be noted that there exists a tradeoff in the optimization of these geometrical offsets with respect to the spot size. From the formula, it is apparent that increasing the spot size will reduce the impact of a transverse offset. However, increasing the spot size will also magnify the impact of any angular misalignment.

For modern fiber connectors, the fiber end is held concentric in a precise ferrule. Two ferrules are aligned in a connector coupling through the use of a common split sleeve. With current manufacturing tolerances on optical fiber, offsets in core alignment are below 1 μm. Main concerns arise around the nature of contact between the two endfaces and the presence of cavity reflections. It can be shown that as two fiber faces approach, standing wave interference will occur in the air gap. This leads to a periodic variation of both the return loss and the insertion loss. At a wavelength of 1300 nm, this variation has a period of about 0.65 μm. Optical return loss can vary between an infinite amount and about 9 dB. Insertion loss will vary between 0 and about 0.6 dB. For flat-faced connections, it is impossible to eliminate this interference, and thus the connector will have poor repeatability. One approach has been the use of a polished end with a slight radius, producing physical contact (PC) to improve the situation. To minimize reflections, angled physical contact (APC) can be used in the polishing procedure. However, angled connectors may exhibit higher insertion loss unless two matched angled connectors are used. Therefore, one has to choose between high repeatability and low insertion loss (with finite return loss) or finite insertion loss (with low reflections).

For fusion splicing, any air gap is usually eliminated by the bonding of the two fiber ends. For mechanical splices, index-matching material is used to fill the gap. Thus in any type of splicing, core alignment is the dominant concern. Again, due to the precision of fiber manufacturing, geometrical profile alignment of two fiber ends typically leads to insertion losses less than 0.05 dB in fusion splicing. Reflections due to interference are not present in a fused splice. However, mechanical splices will still exhibit cavity effects depending on the quality of the cleaves and the index matching.

4.7 SUMMARY

In summary, the optical coupling from one component to another typically involves the efficiency of transfer (insertion loss) and the efficiency of reflection (return loss). For single-mode devices, the geometrical scales involved are such that ray optics is not useful and the wavefront nature of the beams must be considered. In most cases, the optical modes of devices (fibers, lasers, waveguides) may be approximated by a Gaussian beam, and all coupling efficiencies can be viewed in the context of an overlap integral. In situations where the modes can be well matched between two devices, the major problem will be optical alignment. However, with existing tolerances on fiber dimensions, mechanical registration is often sufficient. In situations where the modes are not well matched, both geometrical mismatch and wavefront mismatch will be of concern.

CHAPTER FIVE

Semiconductor Laser and Optical Amplifier Packaging

SCOTT A. MERRITT, K. MOBARHAN, R. WHALEY,
S. FOX, and MARIO DAGENAIS

Reliable, reproducible, high-yield packaging technologies are essential for meeting the cost, performance, and service objectives in systems which utilize optoelectronic components. The specifications of the optoelectronic package include requirements on device reliability, speed, beam quality, coupling efficiency, and fiber alignment retention.

A number of thermal, mechanical, optical, and electrical factors complicate the effort to package semiconductor lasers and semiconductor optical amplifiers (SOAs). First, the threshold currents of semiconductor lasers and the gains available from semiconductor optical amplifiers are quite temperature-sensitive. For example, a semiconductor laser's temperature sensitivity is exponential:

$$I_{\text{threshold}}(T) = I_{\text{threshold}}(T_{\text{ref}}) \exp\left\{\frac{T - T_{\text{ref}}}{T_0}\right\},$$

where T is the active region temperature, T_{ref} is a reference temperature, and T_0 is an empirically determined "characteristic temperature" parameter. Second, semiconductor lasers and SOAs generate large heat fluxes ($4\,\text{kW/cm}^2$ or more in high-power devices) which cause an excessive temperature rise in the active region [YFL93] due to the poor thermal conductivity of compound semiconductors (Table 5.1). Third, these devices are quite sensitive to stress-induced degradation. This makes low-stress die attach techniques essential for optoelectronics. Fourth, the performance of SOAs is strongly dependent on the residual modal reflectivity of the Fabry–Perot cavity formed by the device facets. Low modal reflectivities may be achieved by very high-quality antireflection (AR) coatings or by combining good AR coatings with angled facet or window

TABLE 5.1. Thermal Conductivity and CTE for Various Materials Used in Optoelectronics[a]

Material	Thermal Conductivity, W(m°K) (@20–25°C)	Linear CTE (10^{-6}/°K)
Semiconductors		
InP	67	4.56
$In_{0.47}Ga_{0.53}As$	66	5.66
GaAs (doped)	44	6.4
GaAs (undoped)	44	5.8
$Al_{0.5}Ga_{0.5}As$	11	5.8
$(Al_{0.5}Ga_{0.5})_{0.525}In_{0.475}P$	6	—
$Ga_{0.515}In_{0.485}P$	5	—
Solders		
In	81.8–**86**	**29**–33.0
Sn	64.0–73	19.9–**23.5**
80–20 Au–Sn	57.3	16.0
77.2–20–2.8 Sn–In–Ag	**54**	28
60–40 Sn–Pb	**44.0**–50.6	24.7
88–12 Au–Ge	44.4	12.9–**13.3**
52–48 In–Sn	34.0	20.0
97–3 Au–Si	27.2	12.3
5–95 Sn–Pb	23.0–**35**	**28.4**–29.8
Submounts, Heatsinks, and Heatspreaders		
Diamond (Type IIa)	2000	0.8
Diamond (Type IIa)	4500@200 K	—
CVD diamond	1000–1600	2.0
Silver (Ag)	427	19
Copper (Cu)	398	16.5
Gold (Au)	315	14.4
CVD silicon carbide	193–250	2.3–3.7
15–85 Cu–W (CMSH)	240	7.5
4–6–90 Cu–Ni–W	230	5.4
BeO	220–260	6.5–7.3
30–70 Cu–W	201	10.8
Aluminium nitride	170–200	4.3
Tungsten (W)	178	4.5
SILVAR™	153	6.5
10–90 Cu–W	147–209	6.5
Silicon (Si)	125–150	2.6–4.1
Molybdenum (Mo)	115–140	5.4
Nickel (Ni)	90	13
Enclosures and Other Materials		
Kovar	**16**	**5.9**
Stainless steel	**16**	**17.3**
In Var	**12**	**0.9**
Silica (SiO_2)	**1.2**	**0.6**

[a] Values in boldface are from the Indium Corporation of America's report "Mechanical and Physical Properties of Indalloy Specialty Solders"; the thermal conductivity data in boldface are at 85°C.

structure designs. Fifth, the alignment tolerances in most designs are very small because of the small transverse spot sizes of conventional semiconductor lasers and SOAs. Submicron positioning is needed to achieve coupling efficiencies from within 1 dB of optimum when using lensed single-mode fiber or microlenses in these devices. Sixth, the parasitic capacitances and resistances of the package severely constrain the frequency response of high-speed lasers and detectors. Packaged high-speed devices must therefore be designed and tested to ensure their performance at microwave frequencies.

5.1 A BRIEF HISTORY

The early literature on optoelectronic packaging emphasized low thermal impedance die attach methods because the high current densities of homojunction semiconductor lasers required good heat transfer to allow the devices to operate above cryogenic temperatures. The thermal benefits of a narrow active region, epitaxy-side down mounting, and type IIa natural diamond heatspreaders were recognized as early as 1967 [DyD67]. The diamond heatspreader demonstrated a substantial advantage over direct die attach to copper at 200 K due to the very high thermal conductivity of diamond. The thermal benefits of gold heatspreaders on devices mounted epitaxy-side up were also demonstrated repeatedly [JoD75, MBW92].

5.2 MATERIALS FOR LASER DIE BONDING

5.2.1 Solders and Solder Alloys

A variety of solders and solder alloys have been used for optoelectronic die attach: tin [DyD67], tin–lead [TsS75], indium [FIF79], and gold–tin or gold–germanium eutectic alloys [FFH79]. The need to reduce the stress in devices bonded to diamond heatspreaders led to the replacement of tin with indium [HaH73]. Further reduction in stress in indium-bonded GaAs die was made possible by replacing the diamond heatspreader with silicon and silicon carbide submounts [SaY78, MST83].

The thermal and electrical resistance of indium-bonded devices was found to degrade with time and temperature when the gold in the device contact diffused into the indium solder layer [FIF79, FFH79] and formed brittle, thermally resistive intermetallic compounds, principally Au_9In_4 [JaH89] (Table 5.2). Solder joint embrittlement was also observed when tin and tin–lead solders dissolved or "scavenged" large quantities of gold, leading to the formation of the $AuSn_4$ intermetallic. More stable bonds were shown to be obtained with gold eutectic alloys (88–12 wt % Au–Ge or 80–20 wt % Au–Sn) evaporated on the heatsink [Fuj82]. "Multilayer" metallizations of gold–tin solder alloy were also reported in patents awarded in 1993 and 1994 to Katz and Bacon et al. [KLT93, BKL94] (see Table 5.1). Multilayer gold–tin solders were shown to produce

TABLE 5.2. Intermetallics That Cause Embrittlement in Bonds to Gold

Solder Alloy[a]	Intermetallic Phase	Comments
60–40 Sn–Pb	$AuSn_4$	Embrittlement above 4–5 wt % Au
100 Sn	$AuSn_4$	Embrittlement above 10 wt % Au
97.5–3.5 Sn–Ag	$AuSn_4$	Embrittlement above 10 wt % Au
100 In	Au_9In_4	No embrittlement at 13 wt % Au
		Embrittlement at 60 wt % Au
50–50 In–Pb	Au_9In_4	Embrittlement at 30 wt % Au

[a] Solder alloys listed by weight percentage (wt %).

void-free bonds at bonding temperatures less than the eutectic temperature of the 80–20 wt % Au–Sn alloy (280°C) [Wan92].

5.2.2 Diffusion Barriers, Wettable Layers, and Surface Oxides

Diffusion barriers are commonly used for semiconductor laser die attach to prevent the formation of undesirable intermetallic compounds. Tungsten barriers are effective in reducing the diffusion of significant amounts of gold from the die or the heatsink into indium solder [DMT82]. Nickel barriers, preferably 1–5 μm thick, are also used to prevent the rapid diffusion of copper into indium solders [FrM92].

Wettable layers are important for reliable bonding of thin solder layers, since non-wetting and dewetting problems cause solder voids and weak solder joints. Wettable layers also play a key role in solder bump technology, which exploits the solder surface tension for precision alignment of optical components. Nonwetting problems may be caused by contamination of the substrate or solder or by passivation with surface oxides such as SnO_2 and In_2O_3 or sulfides prior to soldering. Dewetting may occur when the surfaces become passivated during the soldering operation [ASM89]. Thin (0.3-μm) layers of sputtered gold may be used over the diffusion barriers to improve wettability. Nickel–tin wettable layers have also been used to improve wetting of the barrier layer and to prevent tin depletion from gold-tin alloys. When tin is depleted due to the formation of intermetallics, the alloy exhibits premature freezing of the gold-rich portion of the solder and poor solder wetting of the semiconductor die [BKL94]. Tables 5.3 and 5.4 list several published metallizations used for laser and SOA die attach.

The oxides which form on the solder, semiconductor, and heatsink of the submount must be reduced or their formation must be prevented by using oxide-resistant capping layers such as pure gold. Liquid fluxes have traditionally been used to improve wetting by reducing oxides and decreasing the solder surface tension. The corrosivity of most activated liquid fluxes and the residues left by fluxes can reduce reliability and complicate cleaning, especially in small crevices within the optoelectronic package. Several authors have reported success in

TABLE 5.3. Heatsink/Heatspreader Metallizations

Heatsink or Submount	Adhesion Layer	Diffusion Barrier Layer	Transition Layer or "Auxiliary" Layer	Wettable Layer	Solder Layers	References
Silicon, ceramic, or CVD diamond	100 nm Ti	300 nm Ni (or W, Cr, Ru, Mo)			3.2 μm of 75–25 wt % Au–Sn (gold–capped "multilayer")	[KLT93]
Copper	100 nm Ti	50 nm W (or Cr, Ru, Mo) (sputtered)	0.3–2 μm Ni_3Sn_4 & W (cosputtered barrier + wettable layers)	Ni_3Sn_4 (sputtered)	3.0 μm of 75–25 wt % Au–Sn (gold–capped "multilayer")	[BKL94]
Copper		100 nm W		50 nm Au	1–3 μm In	[Wan92]
Copper		200 nm Ni		50 nm Au	2 μm In	[Mob94]
Copper					2 μm Sn	[MHC95]

TABLE 5.4. Semiconductor Die Metallizations Used for Soldering

Semiconductor Die	Contact Layer	Diffusion Barrier Layer	Wettable Layer	Solder Layer	References
InGaAsP–InP	50–100 nm Ti 50–100 nm Pt 500 nm Au				[KLT93]
InGaAsP–InP	50–100 nm Ti 50–100 nm Pt 500 nm Au				[BKL94]
GaAs		20–100 nm W	50 nm Au or Pt		[Wan92]
InGaAsP–InP (on Si substrate)	Ti–Pt–Au				[Mob94]
AlGaAs–GaAs	Cr–Au			2.5 mm In	[MHC95]

replacing liquid fluxes with acetic acid [BTA93] or formic acid vapor [LeB94] for low-temperature soldering (200–230°C). We note that temperatures above 300°C are required to reduce most common oxides with hydrogen-containing/ forming gases (typically hydrogen–nitrogen or hydrogen–argon mixtures), as seen in Table 5.5. These gases serve only as a "cover gas" for typical die bonding temperatures (200–300°C).

Oxidation of the optoelectronic package can be minimized by reflowing the solder or brazing in vacuum. A vacuum oven removes moisture, undesirable gases such as oxygen, contaminants, and gases partially trapped in the assembly. A slow heating or *in situ* vacuum bake prior to solder reflow or brazing assists by volatilizing some contaminants from the assembly. Pressures of 10^{-2} torr are adequate for some solder reflow tasks, but pressures of 10^{-5} torr are needed for gold–silicon eutectic soldering [ScT91] and brazing [She85].

TABLE 5.5. Temperatures Required to Reduce Some Common Oxides in Hydrogen-Forming Gases

Oxide	Reduction Temperatures in Hydrogen-Forming Gas
InO	130°C
PbO_2	> 320°C
SnO	370°C
SnO_2	400°C
In_2O_3	> 600°C

5.2.3 Heatsinks and Submounts

The heatsinks used for optoelectronic die attach may be copper (preferably oxygen-free high conductivity, OFHC), copper–tungsten (10–90 or 15–85 "CMSH"), silicon, CVD silicon carbide (SiC), silver, or advanced composite materials such as carbon fiber copper and silver-impregnated invar ("SILVAR"^{TM}) [BrC94]. Each of these materials represents a tradeoff in thermal conductivity, die stress caused by differences in the coefficient of thermal expansion (CTE), machinability, lapping characteristics, and polishability. For example, OFHC copper has excellent thermal conductivity, but its CTE is significantly larger than for InP or GaAs. Tungsten-rich copper–tungsten alloys have a good CTE match and reasonable thermal conductivity but are very difficult to machine, lap, and polish. The thermal conductivity of silicon is lower than most other heatsink materials, but the micromachinability of silicon (including the formation of silicon V-grooves) and the ability to integrate electronic functions with optoelectronics offers tremendous flexibility for opto-electronic packaging.

Optoelectronic dies are often bonded to submounts rather than directly to the heatsink. Submounts are used to spread heat laterally from the active area of the die, reduce the temperature rise at hotspots, and match CTEs. Diamond submounts reduce the average thermal impedance of the die bond and have proven especially useful in reducing the facet temperature of high-power semiconductor lasers. We note that the CTE of CVD diamond ($2 \times 10^{-6}/°C$) is better matched to optoelectronic components (5–$6 \times 10^{-6}/°C$) than type IIa natural diamond (0.8–$1.1 \times 10^{-6}/°C$) and may be preferred in some applications. Aluminum nitride, SILVAR^{TM}, and SILVAR-K^{TM} offer excellent CTE matches and good thermal conductivities.

The heatsink shape may be designed to increase the thermal mass directly beneath the laser or amplifier facets to reduce the temperature rise due to laser facet heating, as shown in Figure 5.1. The beveled heatsink facets in this design are declined 45–60° with respect to vertical to allow for the high beam divergence angles exhibited by laser diodes with small transverse spot sizes. The edges of the mounting surface or "ridge" on these heatsinks are designed to converge at an angle of 1.5 degrees. This "wedged" design accommodates optical amplifiers of high-power semiconductor lasers with slightly different lengths. The heatsink ridge in Figure 5.1 is oriented at an angle of 24° with respect to the major heatsink facets to allow the principal ray of the input and output beams of 7° angled facet semiconductor amplifiers to be oriented at 90° with respect to the major heatsink facets.

The mounting surface of the heatsink must be lapped very flat in order to achieve good heat transfer from the die. Submicron surface flatnesses and "knife-edge" heatsink edges are desirable, especially in high-power devices. The requirement for mirror polishing of the mounting surface depends on the die attach method [Mob94]. We have found that surface roughnesses of 100 nm (Ra) are acceptable for die bonding with the controlled interdiffusion method described in the next section [MHC95].

FIGURE 5.1. Perspective view of angled, beveled, wedged OFHC heatsink for high-power angled facet traveling-wave amplifiers. Reprinted with permission from [MHC95], copyright 1995 IEEE.

5.2.4   Solder Failure Mechanisms

The decision to use a soft solder, such as indium or 60–40 wt % tin–lead, or a hard solder, such as the 80–20 wt % Au–Sn eutectic, should be made after consideration of the strength, solder migration, creep, fatigue, whisker formation, stress, thermal expansion, liquidus temperature, and thermal conductivity of each solder type. Soft, low-melting-temperature solders are used to reduce the stress caused by CTE mismatch between the die and submount or heatsink and the difference between the solder solidus temperature and room temperature. Soft solders exhibit low yield stresses [FrM92] and are subject to the slow, continuous deformation known as "creep." Solder creep reduces the reliability of the optoelectronic package since the alignment of components is not retained. Fatigue failure may also occur in the joint, particularly with tin alloys which do not contain silver. Soft solders composed of tin or tin alloys, such as the tin-rich gold–tin eutectic (10–90 wt % Au–Sn), can migrate from the solder joint through gold or platinum layers on the submount to form nearly pure tin "whiskers" [Fuk91, p. 311]. Tin can also migrate onto the facets of lasers bonded with the tin-rich 10–90 wt % gold–tin eutectic. Lead (Pb) is known to be effective in suppressing the formation of tin whiskers, and joints made with 60–40 wt % tin–lead alloy are not subject to this failure mechanism [MST83]. Whisker formation has also been observed after electromigration of indium solder. Indium solders are also subject to degradation through the formation of indium hydroxide and indium chloride. Hard solders such as 80–20 wt % gold–tin, 88–12 wt % gold–germanium, and 97–3 wt % gold–silicon do not exhibit whisker formation; these solders also exhibit less creep and higher yield strengths but melt at higher temperatures and cause significant stress in the die [Wan92].

5.3 DIE POSITIONING

Semiconductor laser and optical amplifier dies mounted in the epitaxy-side down configuration for high-power operation must be positioned to within several microns of the heatsink or submount edge. For GaAs/AlGaAs and AlGaInP/ GaInP lasers, the die must not be mounted with any overhang since the characteristic thermal length in these material systems is nearly 5 μm [BrE90, DCG94, MBW92]; overhang subjects these devices to facet heating and thermal runaway [Fuk91, p. 311]. The die should be mounted exactly flush with the heatsink edge or underhung slightly if some degradation in beam quality is acceptable.

5.4 THE CONTROLLED SOLDER INTERDIFFUSION METHOD

One method to achieve low thermal resistances and very reproducible specific thermal resistances in semiconductor lasers and optical amplifiers is to interdiffuse a solder layer on the heatsink or submount with a solder having a lower melting temperature on the epitaxial side of the semiconductor wafer (Figure 5.2) [MHC95, CFM95]. This method is intrinsically free of voids in the semiconductor epitaxial-to-solder interface. The controlled interdiffusion method relaxes the requirements on heatsink smoothness and enables the solder volume and reflow characteristics to be precisely controlled.

During the die bond thermal cycle, the surface temperature of the heatsink is raised to the melting point of the epitaxy-side solder but is not allowed to reach the melting point of the solder on the heatsink or submount; this greatly reduces the chance of solder shorting the device. As the heatsink surface temperature approaches the melting point of epitaxy-side solder, it begins to interdiffuse and alloy with the solder on the heatsink or heatspreader. As the interdiffusion region enlarges, the interdiffusion rate of the two metals increases until the solder bond is complete. The final thickness of the bond is a function of the duration of the die bond thermal cycle and any pressure applied to the die.

Figure 5.3 is an SEM micrograph of a solder joint produced with interdiffused

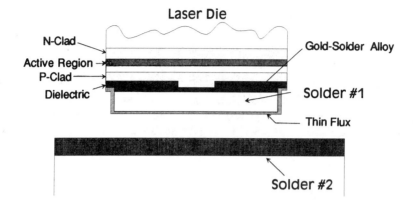

FIGURE 5.2. Die attachment technique which is intrinsically free of epitaxy-side voids. Reprinted with permission from [MHC95], copyright 1995 IEEE.

indium and tin. The device's active region and cladding regions are clearly shown, along with the indium layer, the tin layer, and a partially interdiffused indium–tin layer.

The thermal resistance of bonded lasers and optical amplifiers is determined using the Paoli method [Pao75]. The Paoli method relates the shift in wavelength of a laser diode as a function of applied power to the active region temperature

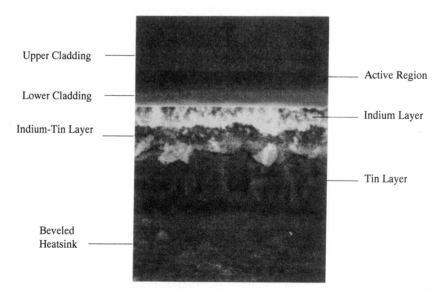

FIGURE 5.3. SEM micrograph of an interdiffused indium-tin solder joint. Reprinted with permission from [MHC95], copyright 1995 IEEE.

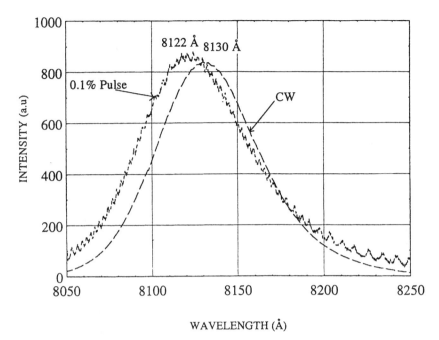

FIGURE 5.4. Wavelength shift versus duty cycle for a GaAs/AlGaAs optical amplifier at
0.1% and 100% duty cycle (CW).   *how to get this data?*

(Figure 5.4). The average power applied to the device is determined from the
applied average electrical power less the average radiated optical power. The
active region temperature is calibrated from the laser diode wavelength shift as a
function of heatsink temperature. For the tapered amplifier in Figure 5.4, the
wavelength shift from 0.1% duty cycle to CW operation is 0.81 nm, the applied
electrical power is 2 W, and the gain shift is 0.293 nm/K; this yields a thermal
resistance of 1.4°C/W. An average specific thermal resistance may be calculated
from the product of the measured thermal resistances and the active area of the
device. We have found that the mean specific thermal resistance of broad-area
lasers is small (4.0×10^{-3} K · cm²/W) and has a small standard deviation
(5.0×10^{-4} K · cm²/W). Nearly identical specific thermal resistances are
obtained on high-power 2-mm-long 810-nm and 970-nm angled-facet tapered
optical amplifiers; these devices consistently achieve thermal resistances of
between 1.4 and 1.6 K/W.

5.5 WIRE BONDING TO GaAs

The fragility of optoelectronic materials reduces the range of parameters
considered safe for wire bonding. In particular, gallium arsenide (GaAs) is not
as hard or as fracture-resistant as silicon and is more susceptible to damage

TABLE 5.6. Selected Mechanical Properties of GaAs and Silicon[a]

Property	GaAs	Silicon
Vickers hardness	6.9 MPa	11.7 MPa
Knoop hardness	590	1015
Young's modulus	84.8 GPa	131
Fracture toughness (energy indent)	0.87 J/m^2	2.1
Compressive force causing electrical defects	15–20 MPa	

[a] Adapted from [Har89a].

during wire bonding. The hardness (i.e., resistance to deformation) and fracture toughness (i.e., the stress or energy required to propagate an existing crack) make GaAs much more subject to cratering. The fracture properties of gallium arsenide depend on defect density; damage to GaAs has been shown to occur with stresses as low as 15–20 MPa (Table 5.6). Thermosonic ball bonding is recommended over ultrasonic wedge bonding to reduce the risk of cratering and damage to the die [Har89a]. The minimum ultrasonic energy should be used when wire bonding, and the work holder temperature should be increased to a safe level to compensate.

5.6 RELIABILITY TESTING

The reliability of lasers and SOAs and the specific degradation mechanisms encountered is strongly dependent on the semiconductor material system, the package materials, packaging methods, and the service environment of the optoelectronic assembly [Fuk91, p. 311]. Lifetest stations such as the one shown in Figure 5.5 and environmental chambers are essential tools for evaluating the impact of these factors on the reliability of the optoelectronic package.

A reliability test station may be used to screen devices, burn in devices to stabilize their characteristics [Fuk91, p. 306], or estimate the life expectancy of the optoelectronic package. In order to limit the test time and the number of devices needed, lifetests are usually performed under "stresses" which increase the degradation rate of the pertinent failure mechanisms. The primary method of stressing semiconductor lasers is to test them under situations of elevated temperatures and/or elevated humidity. This technique, known as "accelerated lifetime testing," is used to estimate device reliability within a reasonable amount of time. The basic method of accelerated testing involves the operation of several samples at a minimum of three separate temperatures. For each test, the laser current is servo-controlled to maintain constant optical power. The operating current lifetest history is recorded at several temperatures for each of several devices. The lifetests are terminated when a predefined failure criterion is reached; the failure criterion is usually taken to be a 50% increase in operating current. Very few devices will actually reach the failure current during an actual lifetest, even at elevated temperatures of 70°C. Instead, the device's mean time to failure (MTTF) is estimated from the rate of increase in current used per kilohour of operation. With these data, the expected lifetime L of the device can be derived via an Arrhenius relation [Fuk91, p. 306]:

$$L = Ce^{E_a/kT},$$

where E_a is an activation energy, k is Boltzmann's constant, and T is the variable denoting device temperature. The activation energy and initial constant are derived from a plot or regression of $\ln L$ versus $1/T$.

Welch et al. [WCS90] have used a variation of the standard accelerated test methodology in which the activation energy is simply assumed and the devices are operated at a single elevated temperature to obtain an estimate of the lifetime characteristic parameter C. This is a useful way to achieve a good rough number for the expected lifetime if the activation energy of the material is approximately known.

A major problem with the standard technique of accelerated testing is that it will not eliminate failure modes that are either temperature-independent or weakly temperature-dependent. Thus, certain wearout mechanisms which can lead to device degradation and subsequent failure will not be fully disclosed under an accelerated test. Gordon et al. [GNH83] proposed a technique known as "purging" whereby all devices are subjected to an initial battery of harsh stresses to weed out those devices which have weak failure mechanisms. These stresses usually involve the operation of uncoated devices at temperatures of 150°C and high operation current. The surviving devices are then tested in the standard accelerated manner but have a significantly reduced amount of oscillation before stabilization, as shown in Figure 5.6. This leads to a faster testing cycle due to the ability to make a time-to-failure measurement at an earlier time.

There are a number of material and processing factors that directly affect the reliability of semiconductor laser devices. Perhaps the most important and difficult to control is the presence of dark line defects (DLDs). The dark line defects are dislocation networks induced by dislocations originally present in the active region [Fuk91, p. 122]. They can be analyzed after device failure using an electron-beam-induced current (EBIC) micrograph. In GaAs/AlGaAs lasers, controlling the material growth conditions to reduce the presence of dislocations in the active region is vital to reducing DLDs. Devices containing DLDs often fail within several hundred hours of lifetime testing and easily succumb to purging techniques. GaAs quantum well lasers have been shown to be quite vulnerable to failure due to DLDs, whereas InGaAs quantum well lasers are significantly more robust [YSD93].

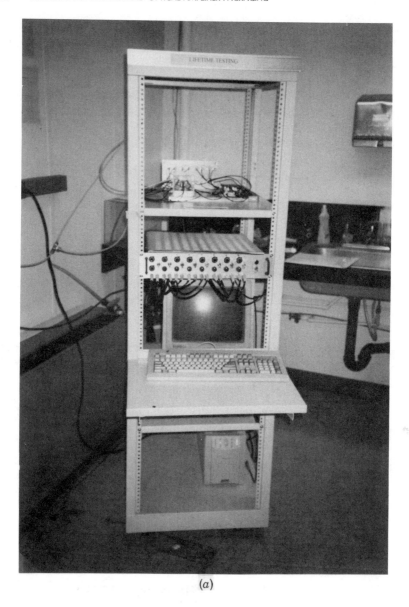

(a)

FIGURE 5.5. (a) Lifetest Station with uninterruptable power supply, data logging computer, drive electronics, and multidevice fixture. (b) Schematic detail of Lifetest Station fixture.

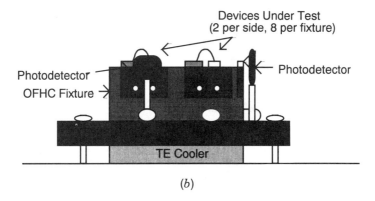

(b)

FIGURE 5.5. (*Continued*)

Another important reliability issue is that of facet treatment. The high surface recombination velocity of the facet–air interface results in a higher temperature at the facet than at the bulk active region. This high temperature reduces the bandgap energy in the region, which in turn increases the absorption of the material, causing further heating of the facet [YFL93]. This is particularly important in AlGaAs lasers, where the exposed Al will readily form its native oxide Al_2O_3 at an elevated temperature. The presence of oxides at the facet will form recombination centers, thus requiring a higher current to be applied to maintain an optical power. However, several methods can be employed to reduce the effects due to facet degradation. If uncoated AlGaAs devices are being tested, they should be hermetically sealed in a dry nitrogen atmosphere to reduce the formation of oxides. Dielectric coatings (i.e., SiO_2 and Al_2O_3) can greatly reduce facet degradation in devices which are not hermetically sealed by eliminating the facet/air interface. Facet degradation may also be reduced by rendering the regions near the facet transparent at the lasing wavelength. These "nonabsorbing mirror" (NAM) regions are made by selectively reducing the bandgap of the active region with a Zn diffusion step [YUK79]; this decreases the lasing wavelength. Since the device facets are not subjected to the Zn diffusion, their bandgap is larger and they are transparent at the lasing wavelength. NAM structures greatly reduce the amount of light absorbed and heat generated at the facets. Large optical cavity structures are also quite beneficial in reducing facet heating [Fuk95].

Mechanical stress is an extremely important factor affecting device reliability. Stress can come from many sources such as die and wire bonding, contact metallization, and cleaving. The degradation of AlGaAs lasers has also been shown to include effects due to Cu migration from submounts [Sim93]. This in turn can result in the formation of dark spot defects (DSDs), which can trigger the formation of DLDs. Thus, the use of barrier metals such as Pt in Ti/Pt/Au metallization on the laser and Ni in Ni/Au on the heatsink is advisable.

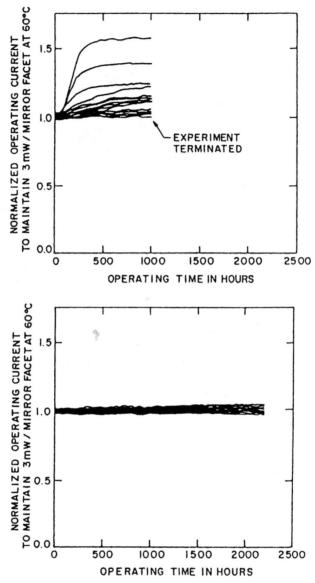

FIGURE 5.6. Degradation behavior of (a) screened device set and (b) screened and purged device set. Reprinted with permission from [GNH83], copyright 1983 IEEE.

5.7 FACET COATINGS

Dielectric coatings are often applied to modify the reflectivity of the facets of semiconductor lasers. High-reflectivity coatings may be applied to both facets to increase the Fabry–Perot cavity effect and decrease the lasing threshold, or they may be applied to a single facet to increase the output power from the opposite facet. Low-reflectivity or antireflection (AR) coatings are essential in order to utilize the optical power gain of semiconductor laser diode materials in semi-conductor optical amplifiers (SOAs). Coating both facets of a semiconductor laser allows it to be used as a broadband optical amplifier. By coating a single facet and using a grating or other external feedback device, one may construct an external cavity laser. AR coatings are applied to many passive optical components, including lenses, optical fiber, waveguide splitters, and directional couplers.

Ultra-high-quality AR coatings are required for broadband semiconductor amplifiers because optical power gain of these devices enhances the gain spectral variation or "ripple" caused by the Fabry–Perot cavity effect. To obtain 1 dB or less of gain ripple with 25 dB of gain, one must realize a facet modal reflectivity of approximately 10^{-4} or better. For a single-layer AR coating, this allows only ± 0.02 index deviation and ± 2-nm thickness deviation from the optimal coating. Thus, *in situ* techniques have been developed to monitor the coating as it is deposited. Unfortunately, materials which are readily deposited as thin films do not exhibit the optimal refractive index for a single-layer coating on GaAs-based devices. SiO_x and SiN_x are used because their refractive index may be varied by controlling their stoichiometry. The refractive index obtained in various coatings also changes due to variations in the rate of deposition and the partial pressure of oxygen or nitrogen in the coating chamber. The presence of impurities in a less-than-ideal vacuum also affects the refractive index of the coating. *In situ* ellipsometric monitoring of thin-film AR coatings overcomes these variables by providing a signal for use in a feedback loop to drive the film index to the target index reproducibly. The optical power output from the device being coated provides an accurate indication that the deposited film is near the optimal thickness and refractive index.

For broadband applications such as tunable external cavity lasers, a single-layer coating does not provide adequately low reflectivity over the entire range. Multilayer coatings and/or angled-facet designs are needed for such applications. *In situ* ellipsometric monitoring allows one to change the outer layer(s) of a multilayer coating to compensate for the variations from the target index and thickness in the underlying layers. The actual thickness and refractive index of each layer are measured and used to adjust the design of the subsequent layers without taking the sample out of the vacuum chamber, which would introduce undesirable oxide layers on the coating.

Measurement of reflectivity spectra of ultra-high-quality AR coatings requires careful measurement and data reduction methods because reflectivities as low as 10^{-7} are achievable. The facet modal reflectivity spectrum is determined by dividing the gain-reflectivity product spectrum of the coated device by the gain spectrum of the uncoated device. These spectra are usually determined using the Hakki–Paoli method or related methods which are synchronized to the peak and valleys of the spontaneous emission spectrum of the device. Our laboratory utilizes the SET method, which orthogonalizes signal and noise

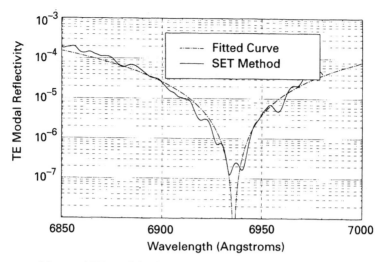

FIGURE 5.7. Measured TE modal reflectivity spectrum for an AR-coated facet computed by the SET method and a theoretical spectrum for a single-layer AR coating. Reprinted with permission from [Mer95], copyright 1995 IEEE.

components, allowing measurement of facet reflectivities down to 10^{-7} (Figure 5.7) [Mer95].

AR-coated semiconductor optical amplifiers have many applications, including optical preamplifiers and high-power optical amplifiers. In free-space optical communications, 980-nm preamplifiers are being used to improve the performance of InGaAs detectors. Precise AR coatings are critical to 980-nm preamplifiers to minimize gain ripple and to suppress lasing of the amplifier, as described above. Coating the facets of such a device to 10^{-4} allows a facet-to-facet gain of approximately 20 dB [Chu93]. AR coatings may also be applied to angled-facet traveling-wave optical amplifiers to reduce the modal reflectivity of the fundamental mode and increase the losses of higher-order modes; this raises the lasing threshold of high-power amplifiers. An AR-coated amplifier with a 7° angled facet structure recently produced output powers of 2.6 W quasi-CW and 1.9 W CW in a diffraction-limited beam with less than 5 mW of input power [CFM95].

Long-wavelength tunable external-cavity lasers also depend heavily on AR coatings. The required facet reflectivity is approximately given by

$$R_{\text{facet}} = R_{\text{ext}}/\alpha^2,$$

where α is the linewidth enhancement factor and R_{ext} is the feedback from the external mirror [Zor94]. For channel substrate planar (CSP) lasers operating in the 820-nm range, α is approximately 3 [IHI92]. If the external feedback is 20% ($R_{\text{ext}} = 0.20$), then the required facet reflectivity is approximately 2%, a value which is easily obtained. However, for lasers operating in the 1.3- to 1.55-μm range, the requirements become much more strict due to a rapid increase in the value of α for wavelengths approaching the band edge. (The value of α for InGaAs lasers increases from about 5 at wavelengths near the center of the gain spectrum to as high as 20 near the band edge [Zor94].) For a device with $\alpha = 20$ and 20% external feedback ($R_{ext} = 0.20$), the required reflectivity is approximately 0.0005. A more accurate calculation based on a rate equation analysis shows that the facet reflectivity should be even lower, approximately 1.5×10^{-4} [Zor94]. We have found that it is necessary to use *in situ* ellipsometric monitoring to reproducibly obtain reflectivities as low as 1.5×10^{-4} or less.

Semiconductor lasers with one facet HR-coated and one facet AR-coated for moderate reflectivity are often used for high-power or intermediate-power pump applications. The HR/AR coating combination is used to obtain higher power from the AR-coated facet while not significantly increasing the laser threshold current. HR/AR-coated semiconductor lasers are used to pump erbium-doped fiber amplifiers at 980 nm. Long-wavelength (2.0-μm) applications such as pumping Ho-doped solid-state lasers, tissue welding and surgery, and spectroscopy may also require HR/AR-coated semiconductor lasers. Choi et al. [Cho95] utilized an AR-coated device to obtain the highest power for a ridge waveguide laser emitting at 2 μm.

Finally, AR coatings are important for more complicated active devices such as traveling-wave amplifier beamsplitters and multicore waveguide power amplifiers. The maximum output power reported for an anti-guided power amplifier utilizing a precise, ellipsometrically monitored AR coating exceeded the previously reported results by more than a factor of three. These results were ultimately limited in output power by the quality of the AR coatings [Zmu94].

5.8   CONCLUSIONS

In conclusion, the performance of semiconductor lasers and optical amplifiers is strongly tied to the technology used to package these devices. The optical, electrical, mechanical, and thermal aspects of the overall design must be considered together, and appropriate tradeoffs must be made to obtain a reliable packaged component. The packaging effort should begin by designing the semiconductor lasers and optical amplifiers for packageability at the outset. For example, design for packageability may begin prior to device epitaxy by incorporating beam expanders into the heterostructure. The packageability of the lasers and SOAs is also improved considerably by incorporating lithographically defined solder pads to "pre-tin" the dies prior to cleaving and dicing. Facet coating technology is a critical part of the overall packaging effort. High-performance facet coatings are designed from a detailed analysis of the modal structure of the semiconductor laser and then deposited using *in situ* monitoring to obtain the required process control.

5.9 APPENDIX: MULTILAYER DESIGN OF SOLDER ALLOYS

Consider a multilayer design for a solder alloy of total thickness T consisting of N materials. The thicknesses t_i $(i = 1, \ldots, N)$ of the individual layers are calculated using the densities ρ_i and weight percentages γ_i of the desired alloy. In the case of four constituents ($N = 4$), the layer thicknesses are given by

$$
\begin{bmatrix} t_1 \\ t_2 \\ t_3 \\ t_4 \end{bmatrix} = \begin{bmatrix} \rho_1\left(\dfrac{\gamma_1 - 1}{\gamma_1}\right) & \rho_2 & \rho_3 & \rho_4 \\ \rho_1 & \rho_2\left(\dfrac{\gamma_2 - 1}{\gamma_2}\right) & \rho_3 & \rho_4 \\ \rho_1 & \rho_2 & \rho_3\left(\dfrac{\gamma_3 - 1}{\gamma_3}\right) & \rho_4 \end{bmatrix}^{-1} \begin{bmatrix} T \\ 0 \\ 0 \\ 0 \end{bmatrix}
$$

This equation can be used to design the overall composition of a multilayer metallization and subsequently used (in conjunction with the phase diagram of the alloy [Mas86]) to partition the multilayer design into a low-melting-temperature alloy on the die and a high-melting-temperature alloy on the heatsink or heatspreader. The final alloy composition will be determined by the extent of the interdiffused region.

As an example of an unpartitioned design, let's consider the indium–lead–silver system. The densities of the constituents are 7.31, 11.34, and 10.49, respectively. If we use an indium–lead–silver alloy with the proportions 80, 15, and 5 wt %, respectively, then a 3-μm-thick solder alloy would consist of 2.58 μm of indium, 0.31 μm of lead, and 0.11 μm of silver.

As an example of a partitioned design, consider the gold–tin system. The densities of gold and tin are 19.3 and 7.31, respectively. The composition of the gold–tin alloy would be 85–15 wt % Au–Sn if it were fully alloyed. (Note that this alloy is gold-rich compared to the 80–20 eutectic.) This composition yields overall thicknesses of 2.0 μm for gold and 1.0 μm for tin. We choose to position all of the tin on the die with 80 nm of gold used as wetting and "capping" layers; this comprises a 17–83 wt % Au–Sn alloy with a melting temperature of 240°C. The remainder of the gold (1.92 μm) will be allocated to the submount.

It is important to note that substantial interlayer diffusion may occur in the vacuum chamber during the deposition of certain solder alloy "multilayers" (such as gold–tin). The overall solder composition will be correct, but the thickness of the last "capping layer" should be designed to be thick enough to account for interlayer diffusion and to prevent subsequent oxidation of the solder alloy.

CHAPTER SIX

Detector Packaging

LAURENCE WATKINS

The basic features of a detector package provide optical and electrical connection to the detector chip. The optical connection is one-way, input only. The electrical connections are signal output as well as supply and bias voltages. In addition, the package provides a benign electrical, thermal, optical, and atmospheric environment. In general, the requirements for these features are less critical than those for the laser package except for the electrical environment. The electrical signal out is very small, and thus the noise contributions from the package must be kept to a minimum. Figure 6.1 is a diagram of a basic detector package with a fiber showing these features.

There can be many design variations from that shown in Figure 6.1, including using lenses between the fiber and detector, integrating a lens into the detector, angling the detector or fiber to reduce back-reflections, and so on. In addition, other devices can be included in the package such as filters to select wavelength and electronic amplifiers to improve detectivity and noise performance.

6.1 OPTICAL CHARACTERISTICS AND REQUIREMENTS

6.1.1 Detector Area and Sensitive Regions

Figure 6.1 shows a typical circular active area for the detector. The size of the active area can range from a few tens of microns in diameter up to several hundred microns or more. Since the capacitance of the device is typically proportional to the active area, high-frequency response detectors will have as small an area as possible. The optical and mechanical design of a detector package deals with directing the light from the input fiber or other source onto the active area.

There can be some variation in detector sensitivity as a function of the position of the incident light in the active area. Figure 6.2 shows a sensitivity plot

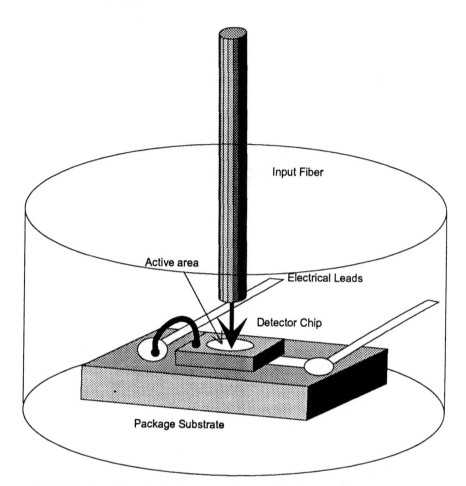

FIGURE 6.1. Schematic of basic detector package with a fiber pigtail showing electrical connections and fiber positioning.

for an InGaAs avalanche photodiode. Also, the detector is often sensitive to light outside the active area unless special techniques have been used to eliminate this. For high-speed detectors, it is important to restrict the light to the active area since the detector speed has be optimized for light incident here. Light incident on other areas of the chip creates electron-hole pairs which take longer to reach the junction region, thus slowing down the speed of response.

6.1.2   Reflection and A/R Coatings

The refractive index of semiconductor detector chip materials is quite high. Table 6.1 gives some values for common detectors. Because of this high refractive index, the reflection of light from the surface would be high—typically 35% for silicon. This would result in significant reduction in sensitivity as well as

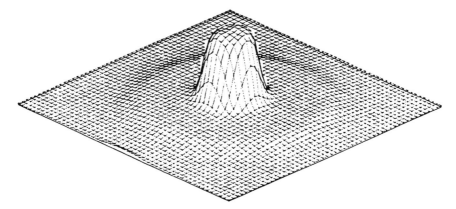

FIGURE 6.2. Two-dimensional plot of the light sensitivity of a detector showing variation over active region and the small response beyond the design detection region. $M = 10$; total scanned area $= 200$ μm $\times 200$ μm; active area diameter $= 40$ μm. Reprinted from data sheet for InGaAs APD 115A/B from Lucent Technologies, copyright 1989 AT&T.

a large back-reflection of the optical signal. Antireflective coatings are normally used to reduce this to 1% or better.

In most applications, the detected signal is low and a reflection of 1% back along the optical path is not a problem. However, in some applications where the detector is used in a monitoring situation, the reflection of a significant portion (-40 dB) of back-reflected light to a laser source will cause noise problems. In these situations, the detector package design must eliminate such back-reflected light by tilting the detector surface to direct the reflection away from the return direction.

6.2 ELECTRICAL PROPERTIES

6.2.1 Equivalent Electrical Circuit and Impedance—PIN, Avalanche, Photoconductor

Figure 6.3 shows the equivalent circuit for a PN or a PIN photodiode. The current generated, I_d, is proportional to the incident light. I_d contains additional

TABLE 6.1. Approximate Refractive Index for Detector Materials

Detector Material	Refractive Index	Wavelength Range
InGaAs	3.5	1000–1670 nm
InGaAsP	dopant-dependent	1100–1600 nm
GaAs	3.34	800–1000 nm
Si	3.41	400–1000 nm
Ge	4.0	600–1600 nm

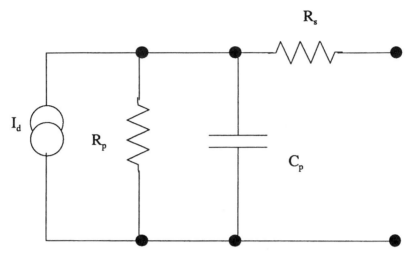

FIGURE 6.3. Equivalent electrical circuit of a PIN detector.

components which are due to the various noise sources. The equivalent circuit for an avalanche photodiode is similar, with the current I_d including a gain factor M from the avalanche gain, which is a function of the applied bias voltage. In a photoconductor, the current generator is replaced by a resistor, the conductivity of which is varied by the incident light.

The photocurrent I_d will usually be specified in units of amps per watt of incident light, A/W, and will typically be 0.4 for silicon and 0.8 for InP detectors. R_p is normally specified in terms of the dark current at a specified bias voltage and operating temperature. Another important specification is the recommended bias and maximum bias voltage which will cause breakdown. InP devices are especially susceptible to electrostatic discharge (ESD) damage, with maximum stress being specified, and so suitable precautions must be taken.

For avalanche photodiodes, additional parameters will be specified such as the bias voltage for recommended gain. The dark current will be given for low gain (primary) and for recommended operating gain, and the latter will be considerably higher.

6.2.2 Noise

There are two primary sources of noise in photodetectors: Johnson noise, due to thermal effects in the resistive components of the device and its circuits, and shot noise or its equivalent, which is due to the quantized nature of electrooptic interactions. For an avalanche device, there are the added random effects of the gain process.

In semiconductor devices, noise is usually given in terms of noise current,

$$\overline{\delta i}^2 = 2eiM^{2+x}\Delta f + \frac{4kT\Delta f}{R},$$

where i includes signal and dark currents, e is electron charge, M is avalanche gain (x depends on APD characteristics, etc.), Δf is frequency bandwidth, k is Boltzmann's constant, T is in Kelvins, and R is the total circuit resistance at temperature T.

For photoconductor devices (including effects of charge amplification), the noise current is given by

$$\overline{\delta i}^2 = \frac{4ei(\tau_0/\tau_d)\Delta f}{1 + 4\pi^2 f^2 \tau_0^2} + \frac{4kT\Delta f}{R}.$$

The first term is analogous to shot noise but includes the effects of carrier creation and recombination. τ_0 is the carrier lifetime, τ_d is the drift time for a carrier to go across the photoconductor, and f is the light frequency. In some cases, noise is specified as NEP (noise equivalent power) and is the input light power equivalent to the noise. For an avalanche detector, an excess noise factor is often quoted which is the noise current power compared to that from an ideal current amplifier realizing the same signal gain.

6.2.3 Detectivity

The performance of a detector is often described using the term D^*, detectivity. This term is useful for comparison purposes by normalizing with respect to detector size and/or noise bandwidth. This is written as

$$D^* = \frac{\sqrt{A\Delta f}}{\text{NEP}},$$

where NEP is the noise equivalent power (for signal-to-noise ratio $= 1$) and A is the detector area. The term $D^*(\lambda)$ is used for quoting the result using a single-wavelength light source, and $D^*(T)$ is used for an unfiltered blackbody radiation source.

6.2.4 High-Frequency Issues

Since the signal from a photodetector is a current source, the typical amplifier circuit used with the detector is a transimpedance amplifier which converts the current to a signal voltage output. The primary parameter that affects frequency response is the device capacitance and parasitic capacitances from packaging, and so on. One method of reducing the parasitics is to include an amplifier chip in the same package as the detector. Some packaged devices of this type are now becoming available.

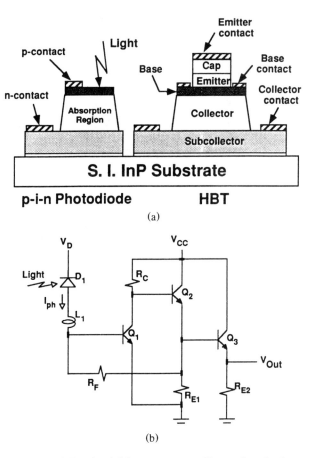

FIGURE 6.4. (a) PIN–HBT epitaxial layer structure illustrating the integration scheme. (b) Circuit diagram of the integrated PIN–HBT photoreceiver. Copyright 1994 IEEE, from [Cow94].

A recent development in InGaAsP detectors has been the integration of an FET transimpedance amplifier on the same substrate as the photodetector. This has resulted in high-bandwidth detection by reducing the parasitics normally encountered in the package and subsequent electronic interconnections. Figure 6.4 shows a more recent report [Cow94] of the integration of an HBT device with a PIN detector; 7.1-GHz bandwidths were reported for this device.

6.3 MECHANICAL PROPERTIES OF DETECTORS

6.3.1 Hermeticity Needs

The need for a hermetic package depends on a number of factors, including detector material, reliability requirements, and the use environment. For low-cost application in benign environments using silicon or even InGaAs/InP detectors, the device will be sealed in plastic packages with flat or lensed windows to input the light. This type of package will permit normal handling and assembly operations and keep dust and dirt which will degrade performance from the surface of the detector. If the detector is going to be used in a dusty environment, the window can be cleaned without damaging the device.

For higher-reliability applications where there is concern about organic contaminants affecting long-term device reliability or when harsh environments can cause condensation and/or freezing of moisture on the internal surfaces, a hermetic package is normally used. Traditionally, data-link, FDDI, and similar types of applications have favored low-cost plastic packaging approaches. In the telecommunications and undersea-type applications, hermetic packaging approaches have been the norm. However, new methods of device passivation and downward cost pressures may change this trend in the future.

6.3.2 Detector Material Strength and Handling Problems

Indium phosphide and, to a lesser extent, gallium arsenide are very fragile and brittle materials—much more so than silicon. This means that, for die bonding and wire bonding operations, special precautions must be taken. In addition, standard equipment which has been designed for bonding to silicon cannot be used or must be modified to greatly reduce the forces it applies to these devices. For example, in die bonding, typical forces applied for InP are tens of grams versus 500 grams or more for silicon. A significant source of failure can be defects caused by wire bonding which will damage the InP and increase dark current to an unacceptable level.

More recent package designs are using a flip-chip die bonding process which makes both electrical contacts to the detector. If the light can be incident from the back surface, then a relatively simple package design, as shown in Figure 6.5, can be used. If the light must be incident from the front, then a means must be provided to bring the light onto the detector's sensitive area on the same side as the electrical contacts. All these die-bonding processes are relatively high-temperature, typically using AuSn solders. This is to allow lower-temperature solders to be used for final package assembly and for mounting the package in a circuit board. These processes have a relatively high cost, and so other materials such as conductive epoxy are being investigated for low-cost packaging. The primary issue to be concerned with is contamination of the detector affecting its reliability.

6.3.3 Thermal Expansion and Heat Conduction Issues

The detector chip itself generates very little heat, even if integrated with an amplifier. The main thermal expansion issues to deal with arise from whatever bonding processes are used during manufacture. Typically, AuSn solder is used to die bond the device, which means the chip can experience 280°C or more

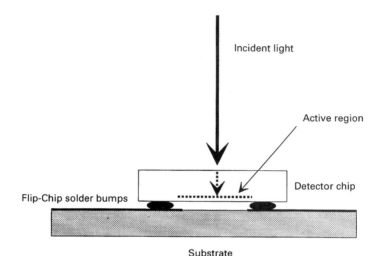

FIGURE 6.5. Diagram of flip-chip bonding showing light incident on backside of detector and transmitting through to active region.

during processing. The chip is usually small in size, and thus stresses from thermal expansion mismatch between the chip and submount are not normally a problem. Methods for alleviating the stress are to use soft solders or to use a soft, thick Au metallization layer. A list of thermal expansion coefficients for various packaging materials is given in Table 6.2.

6.4   WINDOW PACKAGING SCHEMES AND CAPSULES

6.4.1   Detector Capsules as Subcomponents for Packages

The detector chip is usually mounted on a substrate or submount before bonding into a package. As mentioned before, InP and GaAs are quite fragile, and using a

TABLE 6.2. Thermal Expansion Coefficients for Various Detector and Packaging Materials

Material	Thermal Expansion Coefficient (PPM/K)
GaAs	5.8
Cu	16.8
Diamond	1.5–2.8
Silicon	2.7
InP	4.5–4.6
Ge	5.7
Alumina ceramic	6.3–9.1
Au/Sn eutectic solder	16.0

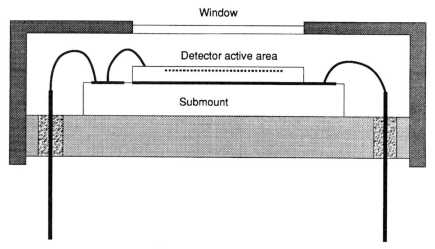

FIGURE 6.6. Simple detector package with window.

submount makes the final packaging operation more reliable and able to achieve higher yield. It also makes it easier for testing and burn-in, if required. Figure 6.6 shows a typical submount arrangement with the detector chip die-bonded to the submount and also shows the electrical connection made with a wirebond. This submount is then bonded into the final package, and electrical contacts are made to the submount using wirebonding or soldering. The submount can be silicon or ceramic. As discussed above, the detector active area can be quite small, down to a few tens of microns, when working with very high frequencies. Because of this, it is important that the alignment procedures and part dimensions, including submount, package, and optics, be designed to achieve the required tolerances when assembled into the final package.

6.4.2   Simple Detector Package with Window

The simplest detector package is a thermooptic can which has a glass window and where the detector is bonded inside facing out the window. Figure 6.6 shows an example design for this. Devices with a flat window are usually large-area devices which can be used in test facilities where custom optics are incorporated to focus the light onto the detector. A simple enhancement is to have a lensed window, often a ball lens, which is aligned coaxially with the detector, as shown in Figure 6.7. This focuses incident collimated light onto the detector window. Figure 6.7 shows the optics for lensed packages showing the numerical aperture $NA = \sin \theta$ and diameter d of the incident light which will be focused onto the detector.

Important concerns are centering of the detector in the package, the distance of the detector surface from the window, and the window thickness. All of these are important when designing the optics and mechanics for the module which

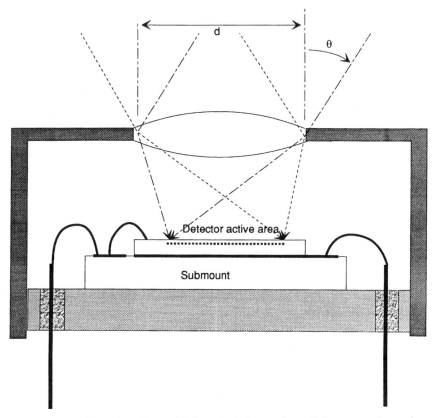

FIGURE 6.7. Enhanced package with lensed window to focus light onto active region.

will contain them. Figure 6.8 shows a selection of packaged detectors of this type from EG&G Co.

6.5   CONNECTORIZED PACKAGES

6.5.1   Various Designs

Connectorized packages are currently used in applications which employ multi-mode fiber, such as data links [Wel87], or for single-mode fiber links operating at lower bit rates and with a modest dynamic range. For high-performance links where high reliability, maximum dynamic range, and stability are required, pigtailed detector packages are preferred. Figure 6.9 shows a detector package design with a connector housing built into the package. There are a multitude of such packages, and some will have signal conditioning ICs built into the package, especially data-link-type applications. Of special note are the FDDI and ESCON transceiver assemblies, which have both detector and source incorporated into a two-fiber connectorized housing. Figure 6.9 shows two

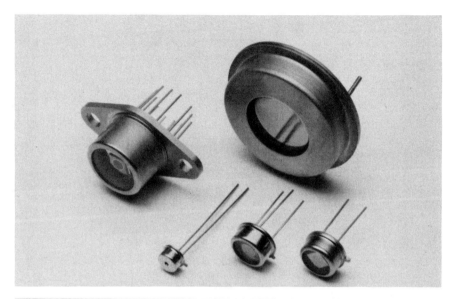

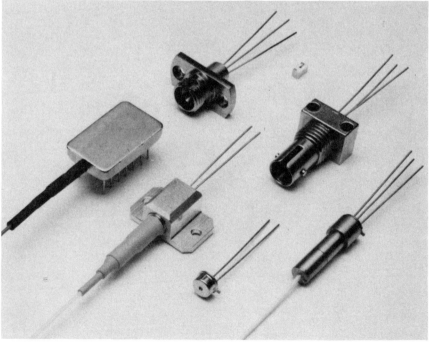

FIGURE 6.8. Examples of various types of detector packages from EG&G. Copyright 1996 EG&G Optoelectronics, Canada.

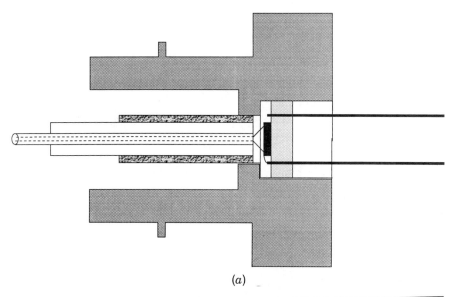

(a)

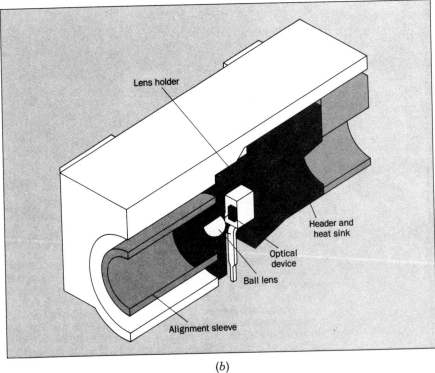

(b)

FIGURE 6.9. (a) Diagram of a detector package with built-in fiber pigtail. (b) Detector package built into a fiber connector for use with connectorized fiber jumper. Copyright 1987 AT&T, from [Wel87].

optical arrangements that can be used in these packages, and Figure 6.8 also shows some examples from EG&G.

6.5.2 Light Coupling Design and Packaging Methods

The configuration of Figure 6.9a is the simplest light coupling design; however, it requires a large detector area to capture all of the light emitted by the fiber, especially multimode fibers whose exit cone angle is typically $\pm 17°$ or more. A better arrangement [Wel87] is shown in Figure 6.9b, which uses a ball lens to image the fiber endface onto the detector. This provides a more efficient light collection and keeps the detector area small for better frequency response.

The connector ferrule and coupling mechanism will maintain the fiber location to within a few microns. This is normally more than adequate to maintain the fiber aligned to the detector window. The window dimension typically ranges from 75 μm (for high frequencies) to several hundred microns (for low-frequency detectors).

6.5.3 Receiver Modules

The detector capsule is often built into a module to provide more functionality. For example, in data links, the signal from the detector goes to an amplifier and then to a decision circuit to provide a digital output of the proper voltages and format, and so on. Figure 6.10 shows a typical design of a data-link receiver package. The connectorized detector module is connected to a printed circuit board, and the whole assembly is put in a box with through hole leads ready for soldering to a circuit board.

Figure 6.11 shows a more modern approach [DBW95] in which a thin-film

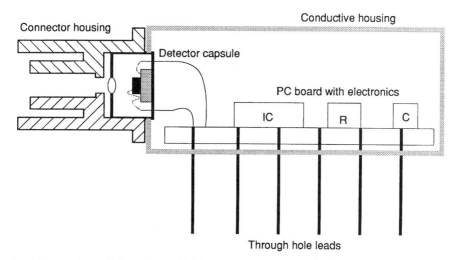

FIGURE 6.10. Data-link receiver which incorporates connectorized detector assembly and electronics on one package.

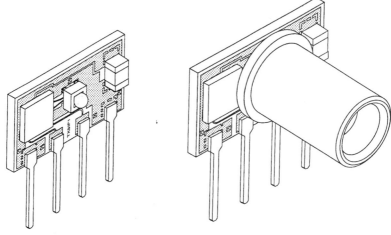

Populated Si Wafer-Board With Lens-Holder Assembly   Ferrule Receptacle Mounted on Wafer-Board

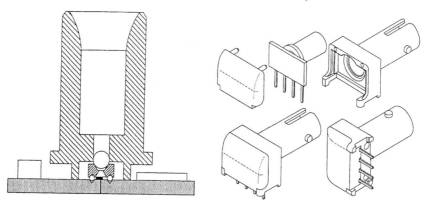

Cross Section/Optical Path   Final Assembly Sequence

FIGURE 6.11. Miniaturized design of data-link using thin-film circuits and compact optoelectronic packaging. Copyright 1995 IEEE, from [DBW95].

circuit is used for mounting the detector chip, receiver electronic chips, and other components. Onto this is mounted a ball lens using a preferentially etched silicon carrier to provide the alignment. This is then assembled into a molded package with the connector ferrule. The molding has the connector form built in.

6.6 FIBER PIGTAILED PACKAGES

6.6.1 Light Coupling Design

The simplest form of coupling from the pigtail fiber to the detector surface is shown in Figure 6.1, where the detector is aligned close to the fiber so that all the light emitted from the fiber endface falls on the detector. The detector area is typically 75 μm. For very high-speed operation it could be smaller; for less demanding applications it could be larger. For highest bit-rate applications, the pigtail is single-mode, and so the light diverging from the 9-μm core can be fully captured by the detector if it is located close enough. The half-angle divergence from a single-mode fiber is typically 8° or so. For example, the fiber must be located less than 270 μm from the fiber for a 75-μm detector.

For lower-performance detectors, which are used for applications which detect signals close to and below 1 Gb/s, a multimode fiber pigtail is often used. This achieves lower loss connection at the fiber pigtail connector because the multimode core is much larger than the incoming single-mode fiber. The same constraints must be observed when aligning and attaching the fiber to the

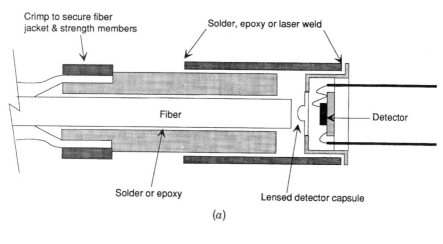

(a)

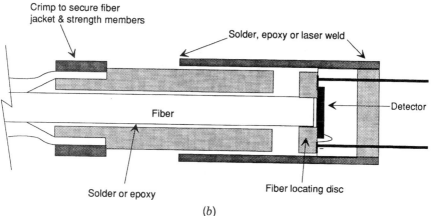

(b)

FIGURE 6.12. Example designs for fiber pigtailed detector packages with and without the use of focusing lenses.

detector. Here the core of the fiber is larger, typically 50 μm, and the numerical aperture is larger so that the emitted beam half-angle divergence is $17°$ or more. This means that the fiber must be much closer to the detector.

6.6.2   Various Designs

Figure 6.12 shows two types of detector package construction. Figure 6.12a is an approach which pigtails the fiber to a detector capsule. This requires a lens between the fiber and the detector to capture the light diverging from the fiber. This approach is the simpler method to achieve a package which is hermetic for the detector.

Figure 6.12b is a package construction which does not require the lens and will achieve the highest coupling efficiency by bringing the detector very close to the fiber. In all these packages, there is an alignment step required to bring the fiber into its correct position for maximum light coupling, and then the bonding process is accomplished to fix it in place. For hermetic packages, soldering or

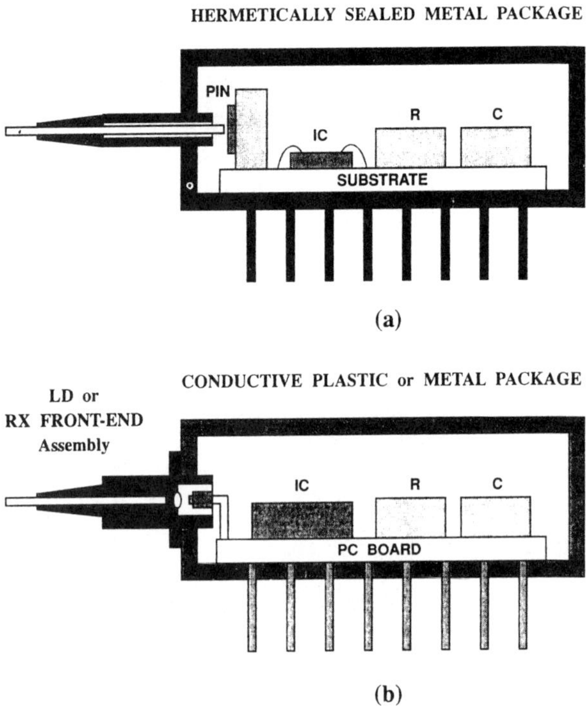

HERMETICALLY SEALED METAL PACKAGE

(a)

CONDUCTIVE PLASTIC or METAL PACKAGE

(b)

FIGURE 6.13. Example designs of pigtailed receiver packages containing detector assembly and electronics. (a) Hybrid and hermetic package. (b) SMT and conductive plastic or low-cost metal package. Copyright 1995 IEEE, from [Muo95].

laser welding is used for bonding. Epoxy is used for less demanding environments for low-cost packages.

6.6.3   Fiber Securing Issues

The fiber pigtail normally consists of the plastic-coated fiber encased in a jacket which can include Kevlar-strength members. This jacket must be securely attached to the detector package so that, when the pigtail is subjected to fiber handling stresses, these are not transmitted to the fiber itself. In addition, it is important to include a bend radius-limiting structure as part of the package so that a sideways pull on the fiber does not cause a sharp bend in the fiber which will stress it to breaking point.

Another issue which has to be considered when assembling the detector package into a module or circuit pack is the layout of the fiber pigtails. Fibers require a minimum bending radius for proper operation; if the fiber is looped with very short radius bends, losses will be induced in the fiber. Typically, fiber bends must be kept to radii larger than 1 inch.

6.7   HYBRID PACKAGES WITH PREAMP ICs AND OTHER ELECTRONICS

6.7.1   Reasons for Hybrids

There are a number of reasons for building modules which contain additional elements to the detector chip. In high-performance applications, very low light

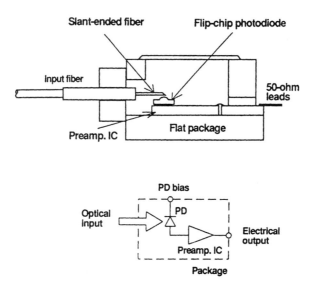

FIGURE 6.14. Example of a fiber pigtail receiver package using flip-chip detector bonding directly onto silicon electronic preamplifier chip. Copyright 1996 IEEE, from [Oik96].

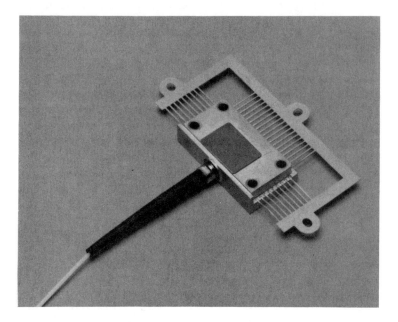

FIGURE 6.15. Photograph of receiver shown in Figure 6.14. Copyright 1996 IEEE, from [Oik96].

intensities and high signal frequencies need to be detected. In this case, it is important to minimize added capacitance and inductance from electrical connections. In addition, the opportunity for picking up noise and spurious signals must be minimized. The best way to achieve this is by putting a preamplifier device as close to the detector as possible.

Special analog circuit design is normally required for communications applications to amplify the small electronic current from the detector and deliver it in the correct digital format ready for further signal processing such as switching, demultiplexing, and so on. It is more efficient and convenient to put all these electronics into the same module with the detector and preamplifier. Thus the system designer can treat this as a black box.

6.7.2   Example Package Designs

Figure 6.13 shows two design approaches [Muo95]. Figure 6.13a uses a hybrid thin-film circuit, and Figure 6.13b uses a small printed circuit board. The thin-film approach can achieve lower capacitance between the amplifier IC and the detector chip. In both cases, the package is sealed in a metal can or a conductive plastic package. The thin-film hybrid package will likely require a hermetically sealed package.

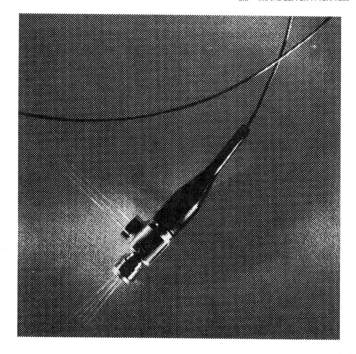

FIGURE 6.16. Fiber pigtailed transceiver package which includes both laser and detector coupled to single fiber. From data sheet from Lucent Technologies for bidirectional laser/detector module. Copyright 1996 AT&T.

Figures 6.14 and 6.15 show a recently reported [Oik94] technique where the detector chip is flip-chip bonded directly to a preamplifier IC. In this way the authors were able to achieve a 10-Gb/s detection rate. This packaging approach uses a novel method of polishing a slant end to the fiber to reflect the light at right angles down onto the detector chip. This avoids the need to mount the detector vertically. Finally, the detector chip uses an integrated lens formed onto the surface of the active area to focus the diverging light from the fiber onto the small active area of the device.

6.8 TRANSCEIVER PACKAGES

6.8.1 Optical and Electrical Crosstalk Problems and Solutions

Most communication applications for fiber optics require two-way communications. Normally this is achieved with two separate fiber links with separate transmitter, receiver, and fiber link. A recent trend in data links has been to develop a transceiver package to integrate the two-way link source, detector, and associated electronics into one module. For applications which use two fibers, there are usually no problems with optical crosstalk. However, a major design

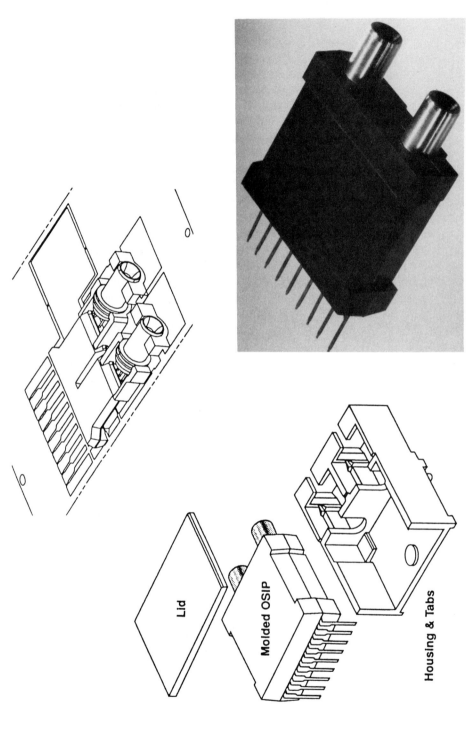

FIGURE 6.17. Example of an FDDI package for two-way communication over a fiber pair showing the assembly sequence. Copyright 1995 IEEE, from [Ani95]

Lid

Molded OSIP

Housing & Tabs

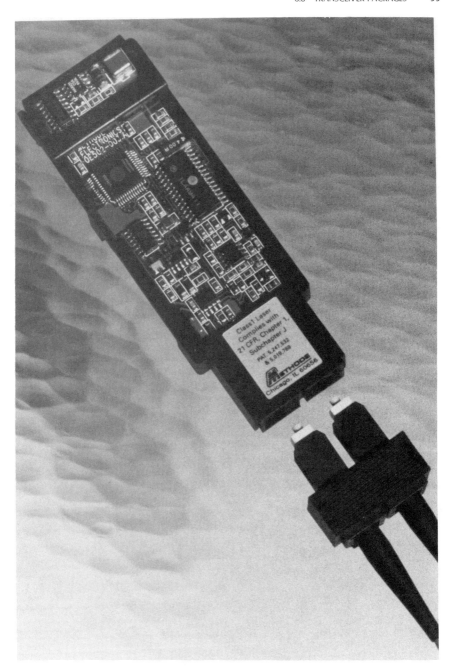

FIGURE 6.18. Example of a fiber channel two-way transceiver module. This contains laser, detector, and all the electronics for signal conditioning and link monitoring and control. From data sheet from Methode Electronics, Inc., copyright 1996 Methode.

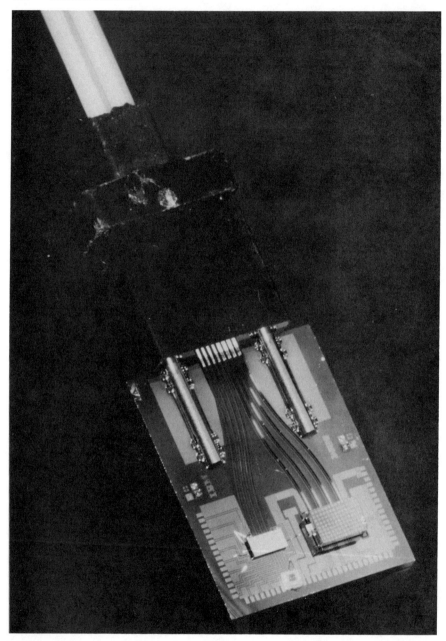

FIGURE 6.19. Photograph of four-channel transceiver module with eight-fiber detachable ribbon connector. Connector retaining fixture has been removed to show module. Copyright 1994 IEEE, from [Jac94].

challenge is to electrically isolate the high-power source drive signals from the low-power signals from the detector. Extensive shielding placed between the detector and its circuits and the laser drive electronics is required to accomplish this.

A different scheme to achieve two-way communication is to do so over one fiber. This requires a splitter to connect both the source and the detector to the fiber. If the light going in both directions is the same wavelength, then the splitter is a 3-dB splitter with the associated total of 6 dB added loss in the link. If the light is of different wavelengths, then this loss is much less and is the sum of the two splitter excess losses. Figure 6.16 shows a bidirectional package from AT&T which contains a detector, a splitter, and a laser source to provide a bidirectional module for two-way optical communication over a single-mode fiber. Optical crosstalk is important for this type of link. Light from the source can be scattered back from the splitter and the fiber connectors into the return path detector, creating noise.

6.8.2 Various Example Designs

There are a couple of standard duplex data-link systems available now, namely FDDI and ESCON. Figure 6.17 shows a packaging scheme recently reported [Ani95] by AT&T for an FDDI transceiver package. It uses capsules for the source and detector which are built into a plastic molded lead frame with the electronics. This is then assembled into the form of the standard FDDI connector. Figure 6.18 shows a fully integrated duplex fiber transceiver for fiber channel applications. This is designed for short-link communications over multimode fiber at 1 Gb/s and contains electronics for all the communication functions, including signal processing and link monitoring and diagnostics. These concepts are being extended to more than one link. Figure 6.19 shows a recent report [Jac94] of a four-channel transceiver module. This is assembled into a package with an eight-fiber ribbon connector to provide a very compact unit.

6.9 SUMMARY

Lightwave communication is increasingly displacing electrical interconnections in all areas of electronic systems and equipment as the cost of lightwave components comes down. The trends in detector packaging will be an increasing densification through arrays and miniaturized packaging coupled with an aggressively declining price curve. This will be accomplished by integrating detectors and detector arrays with electronics through hybrid packaging as well as full integration of more and more functionality into the detector chip.

CHAPTER SEVEN

Waveguide Technologies

ROBERT F. FEUERSTEIN

7.1 FREE-SPACE AND GUIDED WAVE INTERCONNECTS

Optical waveguides (WGs), regardless of technology, share a number of advantages over electrical interconnections—as well as some difficulties. They provide well-controlled and accurately directed connections using optical transmissions. They do not require impedance matching, as do high-speed electrical interconnections, and thus support extremely high bandwidths. They exhibit optical crosstalk when WGs cross, though this can be extremely small when the crossing angles exceed 6° [JaM93]. They do suffer from loss due to scattering in the imperfect WGs and absorption of the optical power by the guiding materials. They can have lower power dissipation than electrical interconnects for moderate-distance communications [Mil89] of > 1 cm, depending on the particulars of the technologies involved. They also easily support large fanouts, since there is no concern for impedance matching. Of course, optical WGs in the form of glass fibers have been used to demonstrate 10,000-km transmissions without signal regeneration by using optical amplifiers, and they have no equal for high-bit-rate and long-distance communications [BAD91, FDB95].

Free-space interconnects, discussed in Chapter 11 of this tome [Mid93, ToC95], have certain advantages over guided waves. Their capabilities are well suited for complex multistage interconnections useful in telecommunications switching applications [MLM94]. They permit two-dimensional interconnection patterns, whereas WGs have generally been restricted to one dimension. We note, however, that a multiple-layer guided-wave polymer interconnect pattern has been demonstrated [TuA93, MKH94].

Free-space interconnects also can have a high density and eliminate the convoluted path from chips on one board, to the backplane, and then onto the next board and its chips. Thus, they also require little board space, since the signals usually travel perpendicular to the board through the air. Two

10-channel, 2-Gb/s-per-channel free-space interconnects have been demonstrated for board-to-board interconnects across a 1-cm board separation [TRW94].

Where guided waves win over free space depends on the particulars of the system being constructed and is the subject of debate amongst the workers in these fields. We will not settle the debate here but will simply point out that there are cases where free space may well be preferable. For the remainder of this section, we will discuss the basics of WGs, their fabrication, and some waveguide devices.

7.2 WHAT IS AN OPTICAL WAVEGUIDE?

Optical waveguides are dielectric structures where the central material, or core, is surrounded by another material, the cladding, of a lower optical refractive index (Figure 7.1). These structures support electromagnetic waves which are contained or guided in the core region as the wave propagates along the guide [Buc92, GhT89, Mar91, Mic93, Pol95, SnL83, Tam90, Yar91]. These waves are stable as they propagate along the WG, and their energy generally stays mostly within the core region. The waves may propagate in one or many modes depending on the wavelength or frequency of the light, the index difference between the core and cladding regions, and the size of the core region. A WG that supports only one mode is called a single-mode (SM) WG. Alternatively, a WG that supports more than one mode is called a multimode (MM) WG. For example, in standard telecommunications optical fibers, an SM fiber at a wavelength of 1.3 or 1.55 μm has a core region approximately 8–9 μm in diameter (Figure 7.1e). Typical MM telecommunications optical fibers have core diameters of 50 or 62.5 μm (Figure 7.1f) and support thousands of modes. Typical channel WG devices are designed to efficiently couple light from and to SM optical fibers, and thus their core regions are usually close in size to an SM fiber core.

The ideal waveguide has low loss (< 0.2 dB/cm), is easily coupled to optical fibers and laser diodes, can be inexpensively fabricated, is polarization insensitive, has low dispersion (wavelength-dependent propagation speed) in the wavelength regime of interest, and exhibits little environmental dependence. For active devices, we add that the material should have a large electrooptic/thermooptic/acoustooptic effect and matched optical and microwave dielectric constants. These properties are useful for low drive power and high-speed active devices such as modulators and 2×2 directional coupler crossbar switches (Figures 7.2c and 7.2d).

Another way to understand the confining nature of the WG is total internal reflection. In total internal reflection, the light inside the core material of higher refractive index n_1 is reflected at the interface between the core and the cladding of lower refractive index n_2 for sufficiently large angles of incidence. The critical angle is the minimum angle for which total internal reflection occurs and is given by

$$\theta_{\text{crit}} = \sin^{-1}(n_2/n_1),$$

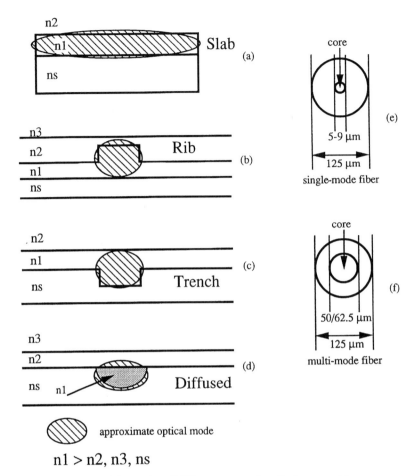

FIGURE 7.1. Examples of different types of optical waveguides.

where the angle is defined with respect to a line perpendicular to the interface (Figure 7.3). One can think of the light propagation in the core as light rays repeatedly being reflected at the core-cladding interface as they move down the WG. Each light ray at a different angle corresponds to a different mode. (Only certain angles or modes are permitted.) Light incident at an angle less than θ_{crit} is partially transmitted into the cladding and results in a loss of power.

While the ray picture above provides a feel for the propagation of light in WGs, understanding the modes is essential to designing any practical devices. The solutions, or modes, of the electromagnetic wave equation will define the characteristics of the WG such as dispersion, give the two-dimensional power distribution of the light, or the so-called mode shape, and lead to expressions for the losses due to waveguide bends and junctions. This information is used to calculate the performance and operation of a multitude of devices such as

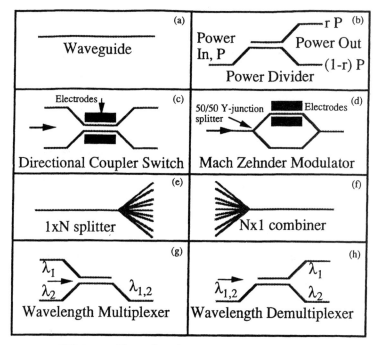

FIGURE 7.2. Examples of channel waveguide devices.

variable ratio power splitters and combiners, wavelength taps and combiners, directional coupler switches, and modulators (Figure 7.2). The first references cited in this section discuss this in great detail.

An important element in determining the mode size is the index difference,

$$\Delta = n_1 - n_2.$$

The larger the index difference, the more mode power is confined to the core of the waveguide. For typical glass optical fibers, $\Delta \leq 0.01$, and perhaps 15% of the optical power extends into the cladding material. Also, Δ plays a role in determining the dimensions for single-mode operation. Generally, the larger the Δ, the smaller the core for an SM waveguide. Δ also affects the numerical aperture (NA) or light-gathering capacity of the waveguide. For step index fibers, which are typical of SM fibers, we have

$$\sin \theta = NA = (2n_1\Delta)^{1/2}.$$

Thus we see that a larger Δ provides more light-gathering capacity, or a larger incident angle θ, into which light can couple into the waveguide. However, generally speaking, most integrated optic devices will be coupled to SM fibers which have NA ≈ 0.1 and cores approximately 8–9 μm in diameter. In coupling

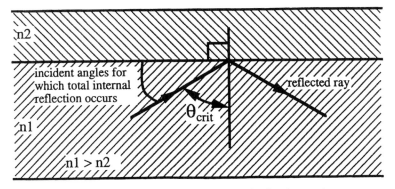

FIGURE 7.3. Definition of total internal reflection angles.

from larger NA to smaller NA WGs, any light power exceeding θ_{max} of the input waveguide will simply be lost, resulting in excess loss. Also, in going from a smaller NA but larger core fiber to a larger NA but smaller core waveguide, light that does not enter the core will be lost, again resulting in excess loss. Thus, there is a tradeoff in designing waveguides between their size and their NA and the loss involved in coupling to the system.

We also note that, generally speaking, metal directly atop the core region results in excessive optical power loss. This is due to the coupling of the optical mode, an electromagnetic wave, into currents in the metal. This effect can be reduced by placing a thin buffer layer of index lower than the core between the metal and the core [CWK91] or by designing the devices so that the electrodes do not sit atop the core regions, as indicated in Figures 7.2c and 7.2d.

7.3 TYPES OF OPTICAL WAVEGUIDES

Optical waveguides come in many different forms, as shown in Figure 7.1. The simplest of these is the slab WG, which only confines the light in one dimension (Figure 7.1a). Channel WGs, on the other hand, confine the light in two dimensions. Also note that the cladding material (index n_2) need not be the same material everywhere as long as its index is lower than that of the core region. However, differing index cladding materials may result in an asymmetric mode shape. The figure indicates an approximate mode shape assuming constant index cladding.

Channel waveguides are currently widely used. Many semiconductor laser diodes use a rib structure (Figure 7.1b) fabricated through multiple complex growth, photolithography, and etching steps. These devices are designed so that the optical gain region of the laser overlaps part of the mode shape of the waveguide.

Another example of an optical waveguide is optical fiber. To prepare glass fibers requires two separate steps. First, a cylindrically shaped glass material is grown and collapsed into a "preform" with a higher-index central region surrounded by lower-index outer cladding. Then it is melted at one end and the glass is "pulled" at a carefully controlled rate though a precise hole to form the fiber [Gow93]. Millions of kilometers of optical fiber are in use throughout the world, with laser diode or light-emitting diode light sources, providing high-quality and high-reliability communications systems. Optical fibers are also melted together to form power splitters and combiners, as well as mechanically actuated 2 × 2 crossbar switches [Opt96].

The selection of a particular WG geometry from those shown in Figure 7.1 depends on the desired device characteristics, including cost, and the materials and processing that can best meet those needs. Most commercial waveguide devices (excluding optical fiber and laser diodes) are formed by diffusion of an impurity into glass or LiNbO$_3$ (Figure 7.1d). Glass is mostly used for passive devices, although thermooptic 8 × 8 switch arrays have been demonstrated with roughly 1-ms switching times [OKO94]. For higher operating speed, the electro-optic LiNbO$_3$ is used for active devices such as optical modulators and switches (Figures 7.2c and 7.2d) [Tam90].

7.4   WAVEGUIDE FABRICATION TECHNOLOGY

In this section we discuss the processing involved in forming optical waveguides on planar substrates. All techniques require photolithography, since the waveguide widths are usually on the order of a few microns. There are basically three approaches to waveguide formation: diffusion, photobleaching, and deposition and etching. Simplified process steps are shown for each of these approaches in Figures 7.4a, 7.4b, and 7.4c, respectively.

Diffusion has commonly been used for forming WGs on planar crystalline (excluding semiconductors) or glass substrate materials. A simplified process diagram for titanium indiffusion in LiNbO$_3$ is shown in Figure 7.4a. A thin film of Ti ($\approx 0.1\,\mu$m) is deposited on the substrate, and then photolithographic techniques are used to define the waveguiding region. Following the photoresist pattern definition, the excess metal is etched away, leaving metal where the waveguide is required. The Ti is the impurity to be diffused.

Alternatively, one can pattern a deposited metal film as the mask for a diffusion step. In that case, a window is opened in the metal where one wishes the waveguides to form. This case is typical of silver or salt ion exchange in glass or hydrogen ion exchange in LiNbO$_3$. In either case, the impurity raises the optical index of the host material. Najafi [Naj92, Naj94] and Tamir [Tam90] cover these techniques in detail.

Certain materials undergo chemical changes when exposed to ultraviolet (UV) and even visible light, such as photoresist. Some of these photosensitive materials can also be used as waveguide materials. Photobleaching typically involves exposing a photosensitive material (usually a polymer) that was spun or deposited on a glass or silicon substrate wafer through a mask to a high-power

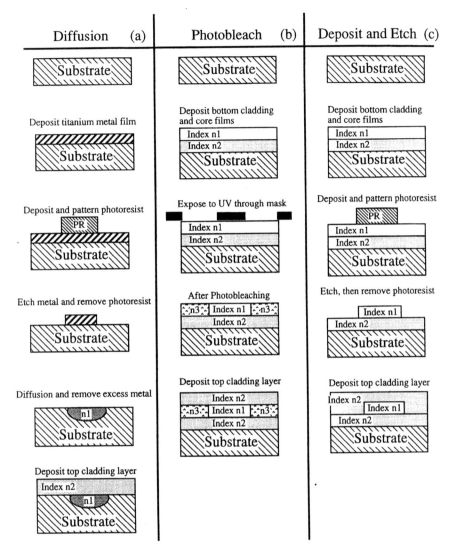

Index n1 > n2, n3, ns

FIGURE 7.4. Simplified process steps for forming optical waveguides.

UV light such as a mercury lamp (significant power at wavelengths shorter than 400 nm). The exposed material undergoes chemical changes that result in a lower optical index [DSM90, Hor92, ITD95, MLF95, ZHK94]. The waveguiding region is the unexposed material. The simplified process steps are shown in Figure 7.4b. This technique would be the simplest way to form optical waveguides on electronics boards or VLSI substrates. Kalluri et al. [KZC96] have recently reported on accomplishing this with EO polymer modulators as well.

Another advantage of the polymer waveguide materials is the potential for multilayer stacks of waveguiding layers which has been demonstrated by Tumolillo et al. [TuA93, MKH94]. This can increase the connection density per centimeter just as it does for electrical interconnects. However, it does of course raise new and difficult packaging issues as well.

The etching technique, usually reactive ion etching (RIE), may be used to form either rib- or trench-type waveguides. Rib waveguides can be formed by etching away the high-index core material, leaving a rib. The core layer is deposited on a lower-index cladding or directly on the substrate. The rib may then be coated with a lower-index cladding material (Figure 7.4c). Alternatively, one can etch away a trench in a lower-index cladding material and then fill it with the higher-index core material, through deposition or growth techniques. One problem with the etching technique is the sidewall roughness of the rib or trench. If the etching technique leaves rough (on the scale of the wavelength of light to be used in the device, typically $\approx 1\,\mu$m) sidewalls, then the scattering losses in this waveguide can be excessive. This can be somewhat alleviated by an annealing treatment that results in smoothing out the residual irregularities.

Many of the standard technologies used in the semiconductor industry have been used for depositing the films. Metals are usually evaporated on where needed. Chemical vapor deposition has been used for cladding and core region depositions. For the waveguide core and cladding, however, some newer techniques beyond chemical vapor deposition are being used. One of these is flame hydrolysis deposition (FHD). This involves depositing silica soot on silicon wafers, which are then heated to form the glass [TJY88]. To differentiate the core and cladding regions, different dopants (GeO_2, TiO_2, B_2O_3, P_2O_5), such as those commonly used in forming optical fibers, are included in the different soot layers. The first layer is a fairly thick ($\approx 15\,\mu$m) cladding layer. This is followed by deposition of the core region and sintering or consolidation of the glass. Then etching is used to form the rib waveguides. Finally, a top cladding layer is deposited by CVD if desired. This technology is currently used by PIRI [PIR96] to produce commercial integrated optic devices. There are two key advantages to this technique:

1. The indices of the core and cladding glass films are the same as those of optical fibers, and thus the mode shapes can be almost identical, resulting in very low coupling losses to optical fiber.
2. The technique is amenable to large-area uniform film depositions, providing some of the largest-area integrated optics substrates available.

Also, since the technique involves fairly high temperatures to reform the glass, the sidewall losses are reduced and losses of 0.01 dB/cm are possible [KOO90]. These substrates have been used to demonstrate thermooptic 2×2 crossbar switches [OKO94]. In addition, erbium has been doped into these waveguides. Erbium is a rare earth element that can be optically pumped and which then provides optical gain in the 1.53- to 1.57-μm region of the spectrum [HKO94].

TABLE 7.1. Comparison of Reported Losses in Various Single-Mode Waveguides

Technology	Material	λ (μm)	Loss (dB/cm)	References
Fiber drawing	Silica fiber	1.55	2.5×10^{-6}	[Gow93]
Titanium indiffusion	LiNbO$_3$	1.3	<0.3	[Thy88]
Ion exchange	LiNbO$_3$	0.63	0.4	[YMT92]
Ion exchange	LiTaO$_3$	0.63	0.3	[YMT92]
Ion exchange	LiNbO$_3$	0.83	0.6 ± 0.2	[VVS91]
K ion exchange	BK-7	1.3	0.25	[MZC89]
Ag ion exchange	Corning 0211 glass	1.3	0.2	[Naj92, p. 66]
Deposit and RIE	polymer	1.3	<0.1	[IYI91]
Deposit and etch	InGaAsP/InP	1.55	2.0	[VOP92]
FHD and etch	Silica	1.3	0.03	[OKO94]
FHD and etch	GeO$_2$-SiO$_2$	1.55	0.01 ± 0.01	[KOO90]

Thus, one of the major pitfalls of integrated optic devices—loss—can be overcome when operating in this wavelength regime. This is very important for telecommunications, since dispersion-shifted optical fibers have a loss and dispersion minimum in this wavelength range [Gow93].

Jet vapor deposition (JVD) is another deposition technique that has recently demonstrated some impressive results [HaS94, ZMS94]. It uses supersonic jets of materials to deposit films. JVD provides very fast, large-area, and low-temperature depositions when compared to other techniques. These advantages offer the promise of inexpensive integrated optics substrates and, more importantly, because of the low deposition temperatures, optical waveguide layers on planar substrates such as circuit boards.

Table 7.1 shows the results of loss measurements in various materials and the technologies for waveguides. We note that optical fibers have the lowest loss by many orders of magnitude.

Hibino et al. [HOH93] have suggested that the combination of the high temperature at which glass fiber is pulled and the elongation it undergoes during the pulling process results in ultrasmooth interfaces between the core and cladding regions and reduced index fluctuations. These in turn give optical fiber its extremely low loss, which approaches the Rayleigh scattering limit of only 0.2 dB/km (2×10^{-6} dB/cm). This is to be compared with the best loss results of $\approx 10^{-2}$ dB/cm in an FHD-deposited and etched planar WG, four orders of magnitude higher loss than fiber with identical materials.

7.5 WAVEGUIDE DEVICES

The range of waveguide-based devices demonstrated to date is truly astounding. They include laser diodes, waveguide-based photodetectors, optical amplifiers, modulators, and power splitters. Wavelength multiplexing as well as demultiplexing devices have also been demonstrated. Directional couplers which can

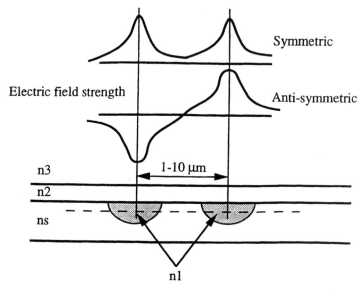

FIGURE 7.5. Schematic of coupled modes and the composite modes that exist when two waveguides are close together. The approximate electric field strength along the dashed line is shown as well.

select individual wavelengths from a set of wavelengths as well as form a crossbar switch have all been used. Figure 7.2 schematically depicts some devices. In addition, acoustically tunable optical wavelength filters have been demonstrated [SBC90, THH94, JBD95]. Various combinations of devices have also been shown. The full range of devices can be seen by examining a few issues of the *IEEE/OSA Journal of Lightwave Technology* or *Photonics Technology Letters*, among other journals. Also see the special issue on integrated optics [JLT88]. Of course, the most common structure is simply a waveguide connecting two points or with some Y-junctions for splitting power off to separate paths.

Many of these devices are based on coupled waveguides. These are waveguides where the optical power is no longer confined inside a single isolated core region but extends over two or more core regions and forms new composite modes. Figure 7.5 shows the two modes of a two-core waveguide, the symmetric and the antisymmetric modes. These two modes propagate with different phase velocities as they go along the guide. They generate an optical interference pattern at the end of the coupled region which is a function of the relative phase difference between them. The interference will result in different optical power at the physically separated (Figure 7.6) output guides, depending on the length of the device. The interference is wavelength-sensitive, and thus these devices can implement wavelength-sensitive devices as shown in Figures 7.2g and 7.2h. The particular behavior of these devices is controlled by choosing the length of the two-waveguide regime and the spacing of the waveguides. The index difference is determined by the technology used. The waveguide size is determined by the

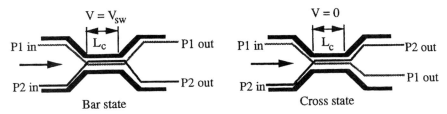

FIGURE 7.6. Directional coupler switch states, Bar and Cross.

requirement for single-mode waveguides. To change the output power behavior of these devices, one only needs to have some means to change the refractive index of the coupled waveguiding region and thus adjust the relative phases of the two modes. This can be done through the electrooptic, acoustooptic, or thermooptic effects, depending on the material and device design.

To make a directional coupler switch, the devices are designed to initially be in the *Cross* state (Figure 7.6). Any perturbation to the waveguides will affect their indices and thus affect the interference pattern at the end of the coupled region. A change in the relative phase of π between the two modes (symmetric and antisymmetric) will result in the switch changing to the *Bar* state. The commonly used means of achieving this phase change are an electric field (in an EO material such as $LiNbO_3$) and temperature changes. Acoustic fields were used to adjust the behavior of coupled-mode devices to form tunable optical filters, as mentioned above. We also note that these devices are sensitive to various environmental perturbations, which can also make them useful as sensors.

Optical modulators for the high-bandwidth SONET OC-192 (10 Gb/s) standards have recently become commercially available [IOC96]. They reduce the dispersive effects seen in fiber links, which use directly modulated laser diodes with their attendant chirp (or wavelength variation during modulation). Modulators typically use the Mach–Zehnder configuration (Figure 7.2d). Here the power is initially split equally into two arms. The phase is modulated in only one arm, and then the electric fields of the two arms are combined and interfere at an output Y-junction combiner.

Table 7.2 lists some waveguide devices, the materials in which waveguides have been fabricated, the response speed, and the material properties useful for active devices.

TABLE 7.2. Properties of Various Optical Waveguide Devices

Active and Passive Integrated Optic Waveguide Devices: Waveguides; polarization splitters; power splitters and combiners; polarization controllers; directional coupler switches; modulators; wavelength demultiplexers and multiplexers.

Materials: Glasses, $LiNbO_3$, $LiTaO_3$, InP, GaAs, Si, Silica, fibers, polymers.

Speed: ms (thermal and strain) $\rightarrow$ μs (acoustic) $\rightarrow$ sub-ns (electrooptic).

Properties: Acoustooptic, thermooptic, electrooptic, birefringence, coupled mode interference.

7.6 OPTICAL ALIGNMENT AND PACKAGING APPROACHES

The use of optical interconnects for interconnecting electronic chips has been investigated for a number of years. It is a major area of research due to the approaching limitations on system designs dictated by the problems inherent in electrical interconnects. Two conferences that were held in the fall of 1995 and the winter of 1996 [MPP95, OIB96] concentrated on this topic. At the SPIE 1996 OIBSA Conference, discussions focused on how the telecommunications and data communications industries can provide the enhanced broadband services envisioned for the National Information Infrastructure. The MPPOI 1995 conference focus "was the possible use of optical interconnections for massively parallel processing systems and their effect on system and algorithm design. Optics offers many benefits for interconnecting large numbers of processing elements but may require us to rethink how we build parallel computer systems and communication networks and how we write applications." These conference proceedings presented new and interesting work to review. Note that there is already an implicit assumption that optics is required for these new high-capacity systems.

In spite of the assumption that optics is needed, we wish to point out a tour de force demonstration using electrical interconnects by Yamanaka et al. [YEG95]. They demonstrated a 320-Gb/s asynchronous transfer mode switch using a four-signal layer copper polyimide interconnect on a 15-layer ceramic power supply substrate, multichip modules, and 96-way flexible printed circuit connectors and cables. This work at NTT pushes the electrical interconnect envelope even further out. However, for the remainder of this section we will discuss optical interconnects.

In order for optically interconnected systems to be realistically manufactured, the packaging issues need to be addressed. As one example of the issues involved, we will discuss an integrated optic chip-based switching node useful for a high-speed distributed switching interconnect [SBR92]. The basic system approach is shown in Figure 7.7a. The inside of the switching node is detailed in Figure 7.7b. This system has the potential to switch 40 Gb/s at each node and constitutes a distributed switch.

The switching node includes four passive optical fiber wavelength division multiplexing and demultiplexing couplers, fiber delay loops, an EO material with six high-speed directional coupler switches, and some optical amplifiers to restore the signal level. There are also arrays of optical receivers and transmitters, and the electronic routing processor, which decides the switch settings, provides the switch drive signals and generates new routing information at 1.3 μm to be transmitted with the outgoing packets. At a higher level (not shown) would be the local computer or communications processor that receives and sends the data.

The integrated optic switching element requires six high-bandwidth electrical connections to drive the directional coupler switches. There are also two six-fiber connectors required to optically couple the switching element to the network and

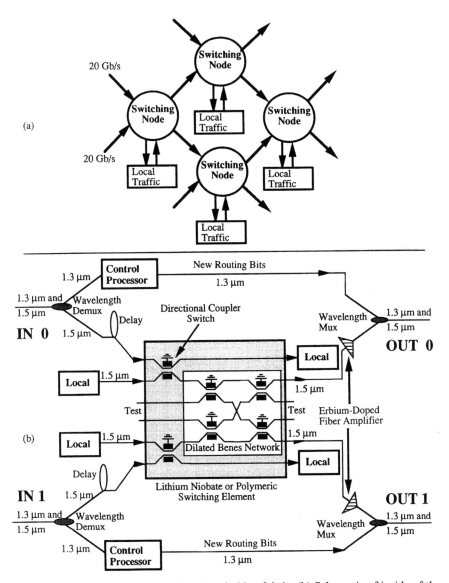

FIGURE 7.7. (a) Overview of optical 2 × 2 switching fabric. (b) Schematic of inside of the switching nodes in (a). The electric connections for the switches are not shown for clarity.

local environment. The packaged device must use standard connectors, both electrical and optical. It must withstand shock, vibration, and temperature excursions appropriate to the environment, and its reliability must be good enough for practical applications. Additionally, of course, one would like its cost to be "reasonable."

The packaging tasks include alignment and fixing of laser diode arrays to

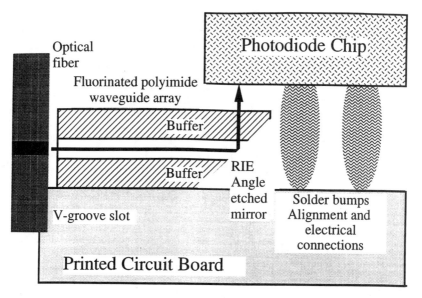

FIGURE 7.8. One approach to coupling fibers to photodetectors with a polymer waveguide interconnect.

single-mode fiber arrays, alignment and fixing of receiver arrays to single-mode fiber arrays (for the local connections and the routing processor), and alignment and fixing of single-mode fiber arrays to the single-mode waveguides in the chip. Since these are single-mode waveguides, the alignment tolerances are on the micrometer scale. Also, there must be low mismatch electrical contacts made from the connector to the electrodes on the chip.

For coupling to photodiode arrays from fibers, one generally needs to change the propagation direction of the light by 90°. Arai et al. [ATK95] used slant-polished fibers to reflect the beam and provide the 90° direction change. Another approach used flip-chip bonding of detector arrays and RIE angle-etched polymer waveguide mirrors, as is described by Koike et al. [KSM96] (Figure 7.8). The fibers are butted up to polymer waveguides using photolithographically defined fiber guides. The polymer waveguides showed less than 0.4-dB/cm loss even after the solder bump heat treatment. Ambrosy et al. [ARH96] and Haugsjaa et al. [HDM96] use a silicon motherboard with an etched face to reflect the beam. These approaches can be used for the connections to the routing processor and the local environment receivers.

Arai et al. [ATK95] describe an active alignment technique to couple a laser diode array to a fiber array. The fibers have hemispherically shaped ends to increase coupling efficiency and alignment tolerance. The fibers are held in precision Si V-grooves formed by anisotropic wet chemical etching of Si. UV-curable cement is used to fix the fiber position once the alignment is optimized. The coupling loss was about −2 dB.

Ambrosy et al. [ARH96] used flip-chip bonding (for $x - y$ alignment) and precise electroplating (for height control) on a silicon motherboard for laser diode arrays. These were aligned to predefined V-grooves which held the optical fibers and achieved a coupling loss of 8.1 ± 0.1 dB. A similar approach for the alignment of laser diodes to single-mode fiber arrays was described by Haugsjaa et al. [HDM96]. They used physical alignment structures etched into the silicon waferboard for passive optical alignment of the laser diode arrays (which were flip-chip bonded to the substrate) to the cleaved fibers in silicon V-grooves. A driver chip was also included on the waferboard. The coupling efficiency was poor at -11.5 dB. The entire transmitter assembly was placed in a commercial flat pack package. In both cases, the coupling efficiency could probably be improved with hemispherical end fibers.

Dautartas et al. [DBW95] describe a self-aligned optical subassembly for alignment of multimode devices. The technique utilizes self-alignment through photolithographically defined and micromachined features in silicon, solder bump self-alignment, precision ferrules, and alignment spheres. This approach can reduce the cost of manufacturing these batch-manufacturable assemblies. Kravitz et al. [KWB96] describe a more complex self-alignment technique, one that is also passive and usable with single-mode fibers. However, it is for coupling to vertical-emitting devices.

The above descriptions are by no means a complete survey of the field but do cover the general approaches currently being considered.

The best general approach described above—namely, automated active alignment with some predefined passive alignment features—can be used for coupling the six-fiber array to and from the switching chip as well as coupling the receiver and laser diode arrays to the fiber arrays. The fiber ends, except for the receiver arrays (the detectors can be fairly large depending on the bit rate, $\geq 70\ \mu$m in diameter), may be hemispherical to improve the coupling efficiencies. The easiest alignment could be achieved with a Si substrate and a buffer-nonlinear polymer-buffer waveguiding region. The Si could be used as the

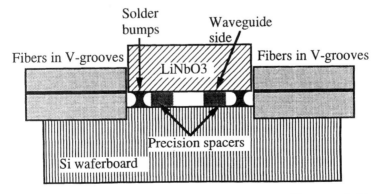

FIGURE 7.9. Technique to physically attach substrate to fiber array holder.

waferboard or motherboard described above to ease the alignment task. If the EO material is LiNbO$_3$, then the fibers must be fixed in a precision V-groove array chip and actively aligned to the LiNbO$_3$ chip. The difficulty then comes with how to fix the two chips relative to one another. Using cement to attach them has been done in the past by Watson et al. [WMB90] with a 16-fiber array at the input and output ends of the LiNbO$_3$ chip. Another approach uses flip-chip bonding with self-alignment achieved through the use of solder bumps. This would be preferred due to its simplicity and micrometer tolerances. Hunziker et al. [HVM94] describe this approach. Figure 7.9 illustrates the basic approach.

This should provide the necessary technology for fabricating these switching nodes. The nodes can also take advantage of volume manufacturing of transmitter and receiver array cards developed for other telecommunications or computer applications. However, there is still much work that needs to be done in developing low-cost, high-precision alignment techniques and reliability testing of the completed integrated optic devices.

CHAPTER EIGHT

A Review of Passive Device Fabrication and Packaging

VENKATA A. BHAGAVATULA

In the following sections, various aspects of pigtailing and packaging of passive components—in particular, fiber and waveguide devices—will be covered. Optoelectronic devices are becoming increasingly important in a number of areas. In particular, for applications involving high-speed processing or communications, optoelectronic devices provide a significant performance edge. A variety of active and passive devices are being developed and fabricated with complex geometries and functions to address these emerging areas of applications. To realize this potential, low cost and high-volume manufacturing capability is essential. As in the case of electronic devices, one of the problem areas has been packaging and interconnection. A number of techniques have been developed to address this issue for electronic devices. A similar effort is being undertaken in the case of optoelectronic systems. This effort can benefit significantly from the earlier efforts in the electronics and optics industry. However, these earlier techniques need to be modified and improved upon, and new techniques have to be developed to address the unique requirements of the optoelectronics industry. In this review, the emphasis will be on component- or chip-level optical packaging. With the emphasis on optical communications in the subscriber loop and data communications environments, the need for high volumes of optical components is increasing. Along with this trend, there is a trend to reduce the cost of the devices. To minimize the installation costs, there is a tendency to install these devices in environments without temperature and humidity control. These factors have a significant impact and put a big burden on the packaging of the devices. In addition, passive optical components are manufactured using a variety of fabrication techniques. Each fabrication technique has its own special requirements for packaging. Similarly, there are a variety of passive devices needed for different applications. A good package is critical to the successful operation of the device under these circumstances. To understand these special requirements, it is necessary to know a little bit about the various device types and the fabrication techniques. A brief overview of these aspects will be covered in the next few pages. This will be followed by a general discussion of the pigtailing and packaging requirements. Testing and reliability of the devices are critical parts of any manufacturing process. A few details of this, along with the reference to Bellcore standards developed to address these issues, will be presented. The discussion on passive component packaging will be concluded with the presentation of an example on a high-reliability coupler for submarine applications.

8.1   PASSIVE DEVICE TYPES

Passive components are used in a variety of applications. A variety of device types have been developed to meet these applications. The requirements for packaging vary based on the device type, fabrication technique, and application. The device types and the fabrication techniques, as well as their impact on packaging, are explained below.

Single-mode and multimode couplers are usually based on one of two phenomena [KaH77, ShG79, BKS80, BHN89]. The first approach is a proximity coupler, shown in Figure 8.1a. In this case, power is coupled between propagating mode fields and undergoes a periodic transfer from one guide to another. The minimum distance for complete power transfer is called the *coupling length* and depends on the waveguide parameters, the physical separation, and the operating wavelength. As a result, these devices typically are wavelength-dependent and may also be dependent on the launch conditions in multimode operation. In devices made by proximity coupling, the coupling phenomenon is quite sensitive to any perturbations that change the waveguide parameters or optical field distributions. For example, bending a proximity coupler can change the split ratio or the wavelength characteristics of the device. For devices of this type, a critical packaging requirement is mechanical rigidity and isolation from external bending forces. This approach is usually implemented by polishing fibers so that the cores can be placed close together or by fusing and tapering the cores so that mode fields will overlap. This process leads to a very thin tapered region that can be very fragile. The light propagating through the coupler would expand in the tapered region and be close to the outer radius of the taper. The absorption properties and the refractive index of the material surrounding this tapered region would affect the coupler performance. In this case, a good package would isolate the tapered region from external perturbations and contact with materials that change their optical characteristics under operating conditions.

The second general method for making couplers, shown in Figure 8.1b, is based on splitting a single guide into two or more additional guides. This is often referred to as a *Y-junction splitter*. Generally, this type of coupler is wavelength-

a)

b)

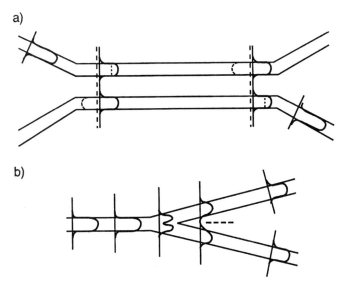

FIGURE 8.1. Coupler fundamentals: (a) Coupling phenomenon in proximity couplers and (b) Y-junction coupler.

independent. Also, the coupling is generally independent of the launch conditions in a multimode coupler. These devices are most often fabricated on planar substrates with pigtail fiber connections. In this particular case, the major concern from a packaging perspective would be the stability of the pigtailing process and its robustness under the required environmental conditions. These are a few examples of how the device type and its fabrication technique would impact the packaging requirements. An overview of the device types is given below.

8.1.1 Coupler Types

A basic coupler type is a $1 \times N$ splitter shown in Figure 8.2a, which splits a single channel into N different output channels. Similarly, this type of coupler can be used to combine N inputs into one output. The N channels are obtained by either a tree-and-branch approach utilizing successive Y-junctions, as shown in Figure 8.2a, or by multiple splitting, as shown in Figure 8.2b. The power is typically split equally between the N channels, and N is usually in the range of 2–32. The upper limit on splitting is determined by the system power available for splitting and the excess loss in the device. Of particular importance is the 1×2 coupler which is used in systems that transmit and receive over the same fiber. The most important configuration for systems being investigated today is the 3-dB coupler—one that splits the input light equally. However, there are cable television applications being developed that make use of a splitting ratio

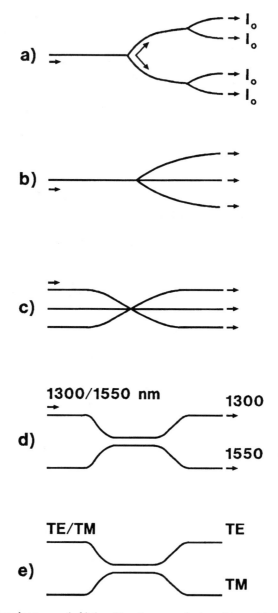

FIGURE 8.2. Coupler types: (a,b) $1 \times N$ or tree coupler/combiner, (c) $N \times N$ coupler, (d) wavelength division multiplexer (WDM), and (e) polarization splitter.

of 5%/95% to broadcast broadband video signals and to receive a low-bandwidth control signal. An extension of the $1 \times N$ coupler is the $M \times N$ coupler shown in Figure 8.2c. In this case, a signal input on any of the M input fibers is split equally to all of the N output fibers. In these cases, without a good design and packaging of the device packaging, the split ratios can vary under different operating conditions.

Another important class of devices is the wavelength division multiplexer–demultiplexer, which is used to combine or to separate signals of different wavelengths onto a fiber. Such devices, shown in Figure 8.2d, are used in systems where different wavelength sources are used to transmit over the same fiber or where one wavelength is used to transmit and another is used to receive. For example, in telecommunication applications, systems are being deployed that transmit both 1310-nm and 1550-nm signals over the same fiber. Narrowband WDM networks are becoming very important in high-bit-rate, long-haul systems with erbium-doped fiber amplifiers (EDFAs). These systems utilize multiple-wavelength transmission with separations of less than 1 nm. In the case of a narrowband WDM device [VeS91, Dra91, CMR92], the wavelength stability of the channels is very critical for proper performance of these systems. In such systems, it may not be possible to eliminate the temperature drift of the channel wavelength, and one may be forced to use a thermoelectric cooler. In such devices, thermal aspects of the packaging become as important as the mechanical and optical issues.

Along with power splitting and WDM applications, polarization is an important parameter to fiber sensors and coherent communications. These couplers, which are usually 1×2, either preserve the input state of the polarization or else separate the input polarization into two orthogonal components, as shown in Figure 8.2e. With these devices, external perturbations that cause stress variations may lead to changes in the device polarization properties and, hence, optical performance.

As indicated in the examples above, the packaging requirements depend on the fabrication technique and also device type. For robust device packaging, the designer needs to know the fundamentals of the device performance and its unique requirements. In the next few sections, a brief overview of the various fabrication techniques will be presented.

8.2 FABRICATION TECHNIQUES

Passive components and couplers, in particular, are fabricated directly from fibers, from fibers with lenses (microoptics), or by using planar techniques [Mil88]. The microoptic approach has been used primarily with multimode fibers and, more recently, in single-mode systems to make components like isolators, tunable filters, and narrowband WDM with two or more channels. In the case of microoptic devices, lenses or GRIN-rod assemblies are used to collimate the light from the input fiber and focus back the collimated light onto the output fiber. Isolator or filter or WDM elements are inserted in the collimated beam. A good example of a microoptic assembly is a fiber-embedded in-line isolator [SYA91]. This paper and the references therein cover various components and, in particular, the assembly process involved in such microoptic devices. With the microoptic components, the main issue is the alignment of a number of discrete parts with submicron tolerances and minimization of the excess loss and back-reflections from the multiple interfaces present in these assemblies. Another issue with these devices is mechanical and vibrational stability. In spite of these issues, a number of microoptic components have been commercially available for the last few years.

8.2.1   Passive Components by Fused Biconic Technique (FBT)

A number of passive devices, in particular evanescent couplers, have been fabricated using fibers [KaH77, ShG79, BKS80]. In this fabrication approach, two or more fibers are fused together and tapered in a controlled fashion using a heat source and motorized stages. This leads to an interaction of the evanescent fields of the two fibers. This technique or improvements on it have been adopted by a number of groups. Using this technique, single-mode couplers with excess losses as low as 0.1 dB have been readily achieved in a 3-dB coupler.

For practical systems, it is important to have the same coupling ratio over the wavelength range of interest. However, as indicated above, proximity coupling is generally a wavelength-dependent phenomenon. Thus, special techniques are needed to make proximity couplers achromatic. An approach for uniform wavelength coupling is an asymmetric design in which one fiber is etched or tapered prior to the fusion process. Power transfer occurs off resonance, ensuring a flatter spectral response. Not only is this technology capable of fabricating low-loss splitters, but the tapering process can be tailored to fabricate a variety of useful devices. The wavelength variation of proximity coupling can be exploited to fabricate wavelength-dependent devices. For example, by controlling the taper, light at 1550 nm can be coupled to the adjoined fiber while 1310-nm light passes through the original fiber. Research and development is also being conducted on a variety of narrowband WDM and other special devices using other special fabrication approaches. For example, index gratings are produced in a fiber by using UV light standing waves to create optical "defects" within the fiber [HFJ78]. Wavelength-selective taps and combiners have been designed utilizing such gratings in order to achieve narrowband components.

Proper coupler packaging is required to ensure good environmental stability. Tapered couplers are very sensitive to any bending in the taper and to any changes in the index of the material surrounding the coupling region. Couplers fabricated with a lower-index glass [TrN86] surrounding the tapered fibers ensure coupling stability, product longevity, and a simplified packaging process. Assembly and packaging details of a high-reliability device made by this approach are given later.

8.2.2   Passive Components by Planar Approach

Planar fabrication techniques offer an alternative to the fused tapered process and the possibility of integrating a number of passive functions on one chip. The planar approach is practiced using a variety of materials and processes. The main fabricational techniques are ion exchange and vapor deposition techniques. The planar techniques have also been applied in a variety of materials. For example, passive optical components have been made in multicomponent glasses, silica on silicon, semiconductors, and polymer materials. Only ion exchange and flame hydrolysis techniques are discussed in more detail here.

The formation of optical waveguides in glass by ion exchange has been explored since the early 1970s [Gia73]. Application of this approach to the fabrication of integrated optical components has also been extensively researched. The geometry of the desired circuit is transferred by lithography onto a thin masking layer deposited on the glass substrate, and the waveguide is formed by thermal diffusion and exchange of the metallic ions of a molten salt bath with ions of the glass structure through the mask opening. To allow for the diffusion and waveguide embedding, the mask openings must be substantially smaller than the final guide dimensions. Typically, silver or thallium is exchanged for sodium and potassium in the glass, and the process can be accelerated by applying an electric field. After this first exchange, a surface index structure with a semicircular crosssection is formed. A second ion exchange with an applied electric field is used to bury the waveguide to reduce the scattering losses and improve the guide geometry. During this exchange, a molten bath of Na^+ and/or K^+ is generally used as the anode. Though the process requires a double ion exchange, very good quality single-mode cylindrical guides can be formed. This enables a good mode-field match between the planar component and an attached fiber.

A number of ion-exchanged multimode and single-mode coupler products have been commercialized [NBJ89]. Multimode tree couplers with N as high as 32, excess loss as low as 0.5 dB (for 1×2), and good port-to-port uniformity of 0.2 dB (for 1×2 couplers) are already available. Such devices are available for all standard multimode types and have achieved low modal dependencies. The current research emphasis is mainly on single-mode devices. With these devices, transmission losses of less than 0.1 dB/cm have been achieved.

Connection loss at the fiber–guide interface can be a major loss factor in integrated optics. With single-mode guides, this comes about because of the core ovality and the mismatch of the mode-field diameters of the guides. However, total coupling losses as low as 0.3 dB/interface have been achieved.

Another approach that is becoming increasingly important for passive component fabrication is the vapor deposition technique. This technique, like ion exchange, follows a planar approach and is also compatible with optical integration and mass production. Although proposals for the application of this technique date back to the early 1970s [KeS74], significant new work has been done only in the 1980s [MKK88, Kaw90]. In this approach, shown schematically in Figure 8.3a, films of glass (mainly doped silica) are deposited on silica or silicon substrates. Film deposition techniques include flame hydrolysis, plasma-enhanced CVD (PECVD), and sputtering techniques. In the flame hydrolysis process, which is similar to optical fiber blank fabrication, silica or doped silica

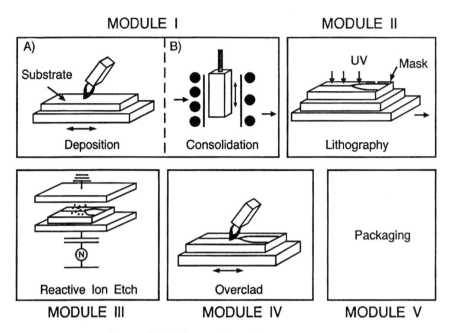

FIGURE 8.3. Planar fabrication technique by vapor deposition process.

soot with the required composition is deposited on a flat substrate to the required thickness. Normal dopants include TiO_2, B_2O_3, and GeO_2. Next, it is consolidated into clear glass at high temperatures. Unlike flame hydrolysis, the PECVD process deposits as clear film at much lower temperatures ($\sim 350°C$). With the PECVD process, the higher-index core layer can be made by P-doped silica or by silicon oxynitrides (SiON). After depositing a thin film, the optical circuit pattern is transferred from a mask onto the glass film using lithographic and plasma-etching methods. An advantage of this technique is that the mask pattern sizes transferred are the same size as the waveguides. This one-to-one pattern transfer relaxes the dimensional tolerances on the mask quite significantly. Even with single-mode devices, feature sizes can be as large as 8–10 μm, which are well within the capability of the lithography process. This approach is well suited for single-mode technology.

Using this technique, various low-loss single-mode devices have been fabricated. Single-mode waveguides with champion research results of less than 0.1 dB/cm for transmission loss and fiber coupling losses of less than 0.05 dB/ interface have been achieved by NTT. $1 \times N$ and $N \times N$ couplers with excess loss in the range of 1–4 dB have been obtained by various groups. In addition, this technology offers the possibility of integrating a number of passive functions on a single chip and the possibility of the hybrid integration of both active and passive components on silicon and other active materials.

With monolithic integration around the corner, planar techniques using semiconductors [Wan88] and polymers [Boo89] to fabricate photonic circuits are coming into prominence. With these techniques, the light-conducting material is a polymer with high light transmission characteristics or a semiconductor like GaAs or InP. The advantage of such polymer materials is the simplicity of thin-film fabrication and of patterning the optical circuits. However, the attenuation characteristics that can be obtained and the thermal and environmental stability of the devices are generally not as good as with glass films. A number of devices, including $1 \times N$ and $N \times N$ couplers, have been manufactured by the polymer technique, and a transmission loss of about 1.0 dB/cm has been obtained. This approach is mainly followed by DuPont, BT&D, and NTT. Because of the nonlinear properties of polymer materials, significant future research activity is expected on these devices. With the potential for easy monolithic optoelectronic integration, much of the work with semiconductor waveguides is limited to active devices, such as modulators, switches, optoelectronic integrated circuits (OEICs), and so on.

Each of these fabrication techniques uses different materials and process temperatures. The package designer needs to understand the impact of these differences to optimize the overall performance. In the next few sections, a general outline of the packaging requirements for the passive devices is given.

8.3 PACKAGING REQUIREMENTS

In general, packaging issues for optoelectronics can be divided into (1) physical size or geometry issues, (2) material issues, (3) mechanical/thermal issues, (4) electrical issues, (5) optical issues, (6) assembly issues, and (7) reliability and testing issues. Many of these issues are considered in great detail in the microelectronics context in a number of other packaging books [TuR89]. In contrast to electronic packaging, optoelectronic devices involve light generation and detection and manipulation of light beams. Because of this additional complication, a number of optical issues need to be considered in the packaging of optoelectronic devices. In this section, the emphasis will be on optical issues as applied to passive optical components.

Manufacturability and low cost are the overriding issues for the data communications and subscriber loop or other consumer-oriented applications. In some of the other applications, the most important requirement is reliability. Examples of such applications are military and undersea applications. In undersea applications, the cost of repair is so prohibitive that it is essential to have very high reliability of the optical components. Also in some of these applications, very high performance over a very broad range of environmental conditions is essential. Also in some applications, as in computer backplane applications or consumer applications, size of the devices is a concern. In such applications, space is at such a premium that compact size of the device is an advantage. All these factors have a significant impact on the device design, technology of fabrication, and the way the device is interfaced with the other devices in the system and the environment. The devices to be interfaced may consist of purely passive devices like fibers, waveguides, filters, attenuators, and polarizers. They could be active devices like modulators, sources, detectors, and so on. The design could also involve the interface between an optical device and an electronic IC chip, as in the case of laser drivers and so on. This variety in device types leads to interfaces or connections that can be categorized as optical, electrical, mechanical, and thermal. In the following few sections, the various approaches developed for optical connections and pigtailing will mainly be discussed.

8.3.1   Optical Issues

A number of techniques have been developed over the years [Che92, Mur88, FTA92] for addressing optical connection issues. Broadly speaking, they can be categorized as waveguide [Mur88] or free space [Che92, Mur88, FTA92] interconnections. In many cases, it is possible and even preferable to use combinations of these two basic techniques.

In guided-wave interconnections, optical connections or transport of the light are accomplished by waveguides or fibers. This approach is quite flexible and could be used in intraboard or board-to-board connections. Generally, fibers are used mainly for distances greater than a few centimeters, whereas the waveguides would be suitable for distances shorter than 10–20 cm. Also, with waveguides it is very difficult to make connections between different boards. Fibers and waveguides can be multimode or single-mode, depending on distance, data rates, and other application requirements. Guided-wave interconnections can be implemented as butt coupling [FTA92], expanded beam, or lensed coupling [Mid79, KYT92]. For high-efficiency coupling between a laser diode and fibers, particularly single-mode fibers, lensed coupling is essential. Depending on the nature of the connection, butt coupling can be fusion splicing or epoxy connection or mateable/demateable connection. An example of a guided-wave connection in waveguides and arrays [Mur88] is shown in Figure 8.4. In these examples, the interconnection is by butt coupling.

As the name suggests, in free-space interconnections, the light interconnection is through free space. The light beam is steered to different locations by lens, mirror, prism array, or hologram. By the very nature of the process, this approach is more suitable for 2-D array-type applications between different boards or planes. Free-space interconnections will be discussed in detail in Chapter 11 of this book.

Specific performance criteria depend on the application and the technology used. However, there are a few generic issues that need to be considered in a majority of cases. Examples of these issues include allowable loss at the connection, back-reflection, optical crosstalk, and so on. These characteristics impact the performance of the overall system. In guided wave interconnections, loss depends on the mode-field mismatch, angular and lateral misalignments, and Fresnel reflections. In addition, if there are any defects or scratches, they also lead to excess loss. The joint loss or coupling loss depends very much on whether the devices are single-mode or multimode. It also depends on whether laser

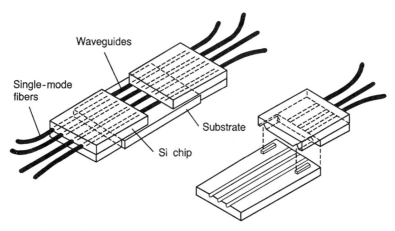

FIGURE 8.4. Array fiber attachment by butt coupling. From [Mur88], copyright 1988 IEEE.

diodes or LEDs are being considered. These factors determine the required alignment tolerances for alignment techniques. Details of the mode-field definitions, the coupling loss calculations based on the mode fields, have already been covered in previous sections and thus will not be covered here. Generally speaking, alignment tolerances for multimode devices are in the few-micron range, whereas they are in the submicron range for assemblies involving laser diodes and single-mode devices [Mur88, Mid79]. This is particularly so in the case of arrays of devices and fibers where this accuracy has to be maintained over a large area or distance. As a comparison, tolerance requirements in many electronic applications are an order of magnitude more relaxed. This is one of the main reasons for the need to develop improved or new techniques. With such stringent requirements, many of the alignment procedures currently used in optoelectronic device assembly involve active power monitoring. Because of tighter tolerances, it is applied much more in applications involving single-mode waveguides and fibers and laser diodes. It is also used in applications where low interconnection loss is essential. However, these techniques are potentially expensive and not easy to scale up to large-scale manufacture. That is the reason for the extensive research efforts for passive techniques that can achieve these stringent tolerances. A short overview of passive alignment techniques and a few examples illustrating the principles of assembly and packaging are given in the following few paragraphs.

8.4 PASSIVE ALIGNMENT AND PACKAGING EXAMPLES FOR OPTICAL CONNECTIONS OR PIGTAILING

Recently, a number of groups have started working on passive alignment techniques for optoelectronic applications. Quite a few of these techniques involve extensions of techniques developed for electronic applications. These improvements and variations are needed to achieve the higher accuracies required in optoelectronics applications, particularly with single-mode devices.

A popular technique involves self-aligning solder-bump, flip-chip technology developed by IBM [Mil69]. It has been used extensively for electrical interconnections with very good success. This is a good technology for electronic devices where a few-micron alignment accuracy is quite sufficient. This technique also allows batch processing. However, for applications in optoelectronics, improved tolerances are essential. A number of techniques to improve the tolerances have been developed [Wal89, PaL91]. In one approach [Wal89], improved accuracy of $\sim 1 \ \mu$m has been achieved. This is achieved with smaller and very well-controlled solder bumps. Reducing the size of solder bumps necessitates more precise placement of the chips. One way of overcoming this difficulty is to use a two-step alignment technique [PaL91] where two different sizes of solder bumps are used. The larger solder bumps achieve coarser alignment, and the smaller solder bumps do the final alignment. Such techniques involve very careful design and process control. This is particularly true with vertical alignment. Another potential limitation of this process is that it involves a large number of complicated steps. Also, the flux needed to be used with this technique may cause contamination of active optoelectronic devices. Due to the high expansion coefficients of the solder bumps compared to glass, there may be significant movement of the devices with temperature variations. In this case, device performance under such conditions may be an issue. A more detailed description of this technique and its applications for optoelectronic devices is given in Section 9.2.

Techniques based on chemical etching of crystalline silicon are widely used for passive alignment and assembly of optoelectronic devices. There are a number of reasons for this approach. Silicon is a very well-studied material and is used extensively in electronics. As a result, it is easy to get very good-quality, large wafers at low cost. The processing techniques are quite well-developed. With crystalline silicon and special wet etching techniques, it is possible to etch V-grooves very precisely with submicron accuracies. Its mechanical, thermal, and electrical properties make it very attractive for these applications. Another attractive feature is that passive waveguides can be patterned on silicon substrates using a number of techniques described in the previous sections. This simplifies the assembly as well as the interconnection issues. A number of groups have used these features for simple pigtailing of devices to more complicated assemblies of sources, detectors, and waveguides. A description of these ideas and packaging examples are given in the following paragraphs.

One variation of the silicon passive alignment technology is the "overlap" self-alignment technique [Mur88] shown in Figure 8.4. This technique allows fabrication of arrays of fibers for interconnection to waveguides or active device arrays. This technique can be used for "quasi-passive" or completely passive alignment. The technique shown in Figure 8.4 allows completely passive alignment of the fiber and waveguide arrays. This is achieved by the interlocking of the ridges on the waveguide substrate with the silicon V-grooves in the fiber array. In principle, the ridges need to be only a few microns high to provide positive locking force in the transverse direction. The ridges are formed in silica deposited over the lithium niobate waveguide substrate. The accuracy of the technique depends on the accuracy of the ridge mask alignment with the waveguides fabricated in lithium niobate. They also depend on the ridge shape reproducibility of the silica etching. In this approach, two sets of silicon V-grooves are used for the dual purpose of fabricating the fiber arrays and also aligning the waveguide chip to the fiber array. The large V-grooves are used to position the fibers and the smaller V-grooves to align the ridges on the waveguide chip. This concept is further extended by Hunziker et al. [HVM93] with improvements in facilitating the electrical connections. The principles of the self-alignment and beam leads beyond the edges of the OEIC chip to provide electrical connections are shown in Figure 8.5. In this case, the same V-grooves are used to position the fibers and align the waveguide chip. The advantage of this approach is that the dimensions of the V-groove can have loose tolerances without affecting the horizontal and vertical alignment of the fiber array and the chip with respect to each other. The fiber array and chip move up or down by the same amount as the V-groove gets smaller or larger. Using this approach, excess losses of ~ 0.5 dB have been achieved.

As indicated above, silicon is well-suited for the purpose of providing an optical bench for hybrid integration and also serving as a substrate for low-loss silica waveguides for lightwave circuits. However, flat substrates with thick waveguide layers deposited on them would pose a problem to etch the V-grooves and other features needed to facilitate hybrid assembly. Also, the various devices that may go on top of the silicon substrate may have their optical axes at different heights from the surface. To handle such situations, an approach using terraced

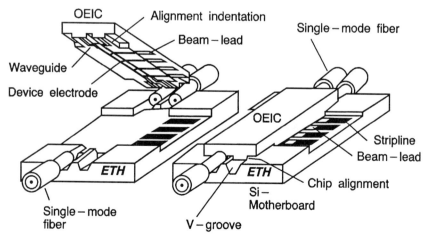

FIGURE 8.5. An example of optically self-aligned OEIC chip on Si motherboard using V-grooves. From [HVM93], copyright 1993 IEEE.

silicon substrate [YTO93] as the platform has been suggested. A schematic of the terraced silicon substrate with waveguides and fibers and optoelectronic chip is shown in Figure 8.6. In this case, a thick undercladding layer is deposited on the silicon substrate with a terraced region. Next, the surface is polished flat so that the thick undercladding is left only in the cavity regions, exposing the silicon terrace in the rest of the areas. On this flat substrate, lightwave circuits are built using conventional techniques. Finally, the top waveguide layers are etched out on the terraces using a process that uses the silicon terrace as an etch stop. The passive alignment structures are fabricated in the terraced silicon areas. Using this approach, waveguide losses as low as those obtained by conventional flame hydrolysis techniques have been achieved.

Another variation of the silicon motherboard approach [Jac92, ANT92] involves the use of stopper pedestals to position the chips. The chips can be laser or detector chips or passive waveguide chips. These stopper pedestals have been proposed to be used with silicon V-grooves to achieve passive alignment with good vertical alignment accuracy. The coupling efficiency obtained by an array of laser diodes coupled to single-mode fibers is in the range of 3–6%. This is comparable to efficiencies obtained by butt coupling with active alignment. The assembly technique involves pushing the chip against these stoppers/ pedestals, and this may be damaging to the chips. The use of standoff pedestals to obtain vertical alignment can be combined with the solder bump technique [ANT92]. This would eliminate the need to push the chip against the pedestal. Also, this would reduce the control requirement on the solder bump size. The results in this case have been with multimode fiber, and the alignment accuracies achieved are in the 1 to 2 μm range. A more detailed description of this work is given in Section 9.2.

In the examples described above, V-grooves fabricated by selective wet etching have been used. In applications where such wet etching is not suitable or crystalline silicon cannot be used, the following technique [GDV91], which

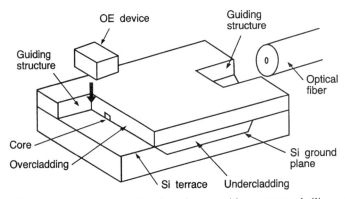

FIGURE 8.6. Schematic structure of silica-based waveguide on terraced silicon substrate. From [YTO93], copyright 1993 IEEE.

uses U-grooves, may be applicable. A schematic of the etched U-groove is shown in Figure 8.7. The advantage of this approach is that this can be done using dry etching techniques, which are generally compatible with other semiconductor processes. Also, this can be implemented with crystalline or amorphous silicon. As shown in Figure 8.7, when used with silica on silicon devices, an undercut is produced due to the higher etch rates for silicon compared to silica. This exposes the silica waveguide for easy butt coupling of the fibers. The disadvantage of this process is that dimensions of the U-groove, particularly the etch depth, are not self-limiting. For low-loss pigtailing, very good process control is needed. Feasibility of losses as low as 0.4 dB has been experimentally verified.

A number of other passive [YaK87] and quasi-passive alignment [CCB91] techniques have also been proposed and tested. The results obtained with many of these techniques are in the 1 to 2-μm range, with a potential for submicron capability with improved process control. Some of these techniques are conceptually simple but involve complicated processing. As indicated in the introduction, there are a number of mechanical, electrical, thermal, and environmental factors that need to be considered for any interconnection scheme to work in practical situations. Hence, the alignment techniques and device designs should take this into account. This is especially the case with laser source coupling to waveguides and fibers. With laser diode sources, the beam diameters are quite small compared to fiber mode-field diameters. To overcome this problem, lenses are used to improve the coupling efficiency. This adds to the cost of the package and reduces its reliability. The ultimate goal is to reduce the cost of assembly without a sacrifice of performance. New techniques may become necessary to achieve that goal. The stringent requirements for alignment can be relaxed with novel improvements in the design of the devices, as indicated in [LRH94]. In this example, spot size of the laser diode is increased to improve the coupling efficiency and alignment tolerances without the need for a separate lens. This is accomplished by a laser structure where the active layer is tapered to facilitate good coupling to the passive waveguide layer underneath it. This twin

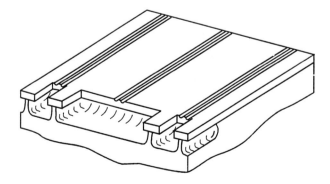

FIGURE 8.7. Schematic of the V-grooves and channel waveguides.

guide structure puts fewer constraints on the laser diode design. In the future, one can expect to see more such optimizations.

Finally, an example of a high-reliability passive component designed for undersea applications [YMG94] is described. This example illustrates the principles of packaging applied to a fiber coupler device. A schematic of the coupler and its package is shown in Figure 8.8. The device uses a patented MultiClad™ process which is a variation of the fused biconic taper technique described previously. This technique consists of a high-silica tube collapsed around the fibers before tapering. This process results in a robust index-controlled, hermetically sealed device that is insensitive to external variations of the index. The packaging had to take into account the mechanical shock and vibration inherent in undersea installations. This was done using a composite wrap material that is very stiff, with near-zero thermal expansion over the temperature range of interest. The composite wrap consists of a unique combination of carbon fiber, phenolic resin, and negative-expansion glass–ceramic powder. This material had to have good adhesion to the silica glass and negligible shrinkage on curing. The high modulus of the wrap provides bend resistance and minimizes any changes in coupling ratio under mechanical loading. This wrapped coupler element is housed in an LCP package to provide additional isolation from external bending forces. The pigtailed fibers are enclosed in loose tubes and protected with silicone boots to provide durability under the pulling forces that can be present during installation.

In addition to proper design, another important aspect of high reliability is testing. Extensive testing of not only the final product but also the individual materials that go into it is very important. A number of testing procedures called

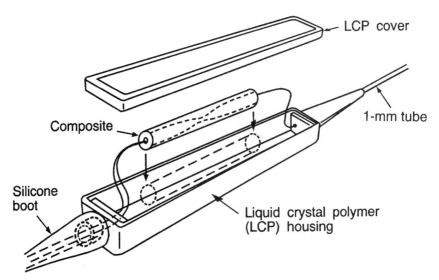

FIGURE 8.8. Packaging example of a high-reliability coupler for undersea applications.

technical advisories have been developed by Bellcore for passive components. For 1 × N splitters, the relevant technical advisory is "Generic Requirements for Fiber Optic Branching Components," GR-1209-CORE. This contains various design and performance criteria and the test procedures.

8.5 SUMMARY

In this section, a review of the fabrication and packaging of passive optical devices is given. A good understanding of the device principles and its fabrication techniques are essential for the development of a robust device. Such a study allows an understanding of the critical performance issues and their packaging requirements. With passive optical devices, one of the main packaging issues is the interconnection issue. Currently, the emphasis is mainly on pigtailing these devices. However, in the future one can expect a trend toward miniature connectorized devices. Presently, the majority of the pigtailing or other interconnections in passive optical components are being done using active alignment. However, to achieve the goal of low-cost devices, it is essential to develop passive alignment techniques that are simple in implementation. A variety of potential technologies and approaches and their performance status have also been covered. These techniques must be applicable not only for passive components but also for active devices. With the need for multiport and array devices, these passive techniques have to be applicable over significant sizes. A number of the techniques developed so far indicate that such a prospect is possible, but more research is needed.

CHAPTER NINE

Array Device Packaging

NAGESH R. BASAVANHALLY and RONALD A. NORDIN

9.1 INTRODUCTION TO PARALLEL INTERCONNECTS

The systems of the future in the computer, telecommunications, and military industries will require high-throughput connections throughout the interconnection hierarchy in order to provide advanced services at competitive rates. Such advanced services include Broadband Integrated Services Digital Network (BISDN), entertainment video, high-definition television (HDTV), video teleconferencing, high-data-rate computer mainframes, Internet access, parallel processing systems, virtual reality techniques, and new military applications. The interconnection hierarchy includes intra- and interframe, shelf, circuit board, and integrated circuit connectivity. End-to-end optical data-link (ODL) products have been successfully used in telecommunication products (e.g., long-span systems and within switching systems) and data networks (e.g., FDDI, Fiber Link, LANs). To meet the needs of future systems, it has been argued [Nor92] that parallel optical data links based on laser array technology are required to meet performance and cost expectations.

Although parallel optical interconnect technology has high potential, its implementation has been slow due to lack of commercially available transceiver array packages. To date, the array packaging is still in the laboratory phase, and the design is mostly limited to buttcoupling of fibers to devices because of its relaxed positional tolerance [Bou94] and to reduce coupling loss variation between different channels [Tak94]. Its success will depend on how well it can compete economically with high-data-rate serial optical interconnect technology as well as parallel electrical technology. The serial optical solutions compete at longer distances (e.g., greater than 50 meters), and the parallel electrical solutions are competitive at shorter distances (e.g., less than 10 meters). Recently, electrical solutions have gained attention due to improvements in coding and compression techniques which lower the required bandwidth and hence raise performance. The window of opportunity for the development of parallel optics depends primarily on implementation cost. The expense involved in optoelectronic packaging, especially for the transmitter packages, can be attributed to alignment requirements. The alignment requirement can vary from submicron ranges for single-mode applications to 5–10 μm for multimode interconnections. Also, the inclusion of optical elements such as ball lens and GRIN lens to improve coupling efficiency translates to an increased sensitivity to misalignment. Though passive alignment technology has the potential to be an inexpensive scheme—such as passive alignment of a laser array to an optical single-mode fiber array [Arm91, Coh91], it is still in its infancy. Hence, passive alignment techniques are predominantly used for multimode fiber applications. It is therefore very important to optimize the package design based on both cost and performance requirements.

Parallel optical interconnection will most likely be achieved using generic components designed to communicate over distances of between 1 meter (m) and 10 kilometers (km). Although the performance requirements depend on a particular system, parallel optical links should in general have (a) very low skew (delay variation within a group), (b) low power consumption, (c) high reliability, (d) bit rates between 150 Mb/s and 1 Gb/s, (e) small circuit-board footprints, (f) small cable cross-sectional area, and (g) most importantly, low cost per interconnection. The key components and technology needed to meet the above requirements are laser diode arrays, detector diode arrays, electronic laser driver and receiver arrays, low skew fiber arrays, optical submount technologies, and optimized alignment and attachment technologies. Optical array technologies are preferred due to their performance (i.e., matched characteristics), lower cost of assembly (i.e., when one device is aligned, the rest are aligned as well), and an overall lower cost (i.e., low incremental optical element cost as well as packaging cost is amortized over all channels). Active array device packaging can be either pigtailed or connectorized type. The current industry trend is toward using a connectorized version [NBH93, Bur94] due to its ease in mounting on circuit boards. However, pigtailing is still used in packages with SM fibers for lack of corresponding connectorization technology. An issue which the parallel optical community must still address for large-volume acceptance is the concern regarding multivendor compatability (with respect to coding, wavelength, fiber, connector, and circuit-board footprint). Hence, standards (e.g., SCI) must be established and accepted for parallel optical modules.

9.2   ONE-DIMENSIONAL ACTIVE DEVICE ARRAYS

9.2.1   Optical Transmitter Sources

The three known optical source devices are

1. Light-emitting diode (LED) arrays
2. Optical modulator arrays
3. Laser diode arrays

The LED arrays [Kae90] are attractive because they are inexpensive, highly reliable, easily extendable to large arrays, and less temperature-sensitive than other optical sources. However, some drawbacks of LEDs are low external power efficiency (a coupled power efficiency of 0.2% with an integral lens), long carrier lifetime compared to lasers, and, significantly, the inability to supply the necessary light power.

Optical modulator arrays are devices which modulate light using the following signals either individually or in combination: electrical signal (e.g., LiNbO$_3$ or InP coupler), optical signal, magnetic signal, and mechanical signal. Usually, the light power source is carried in a fiber-optic cable, and the light is then split inside the data-link package and routed to individual modulators. Examples of optical modulators include LiNbO$_3$ [PaH87], liquid crystal devices (electrooptic effect), deformable mirrors [PaH83], and magnetooptic devices [Ros82]. The self-electrooptic effect device (SEED) [Nov93] is an example of the latter type of latchable modulator.

In recent years, parallel interconnect research has mainly involved the use of laser arrays. Lasers are preferred over LEDs because of (a) their higher external power efficiency, (b) their narrow beam which allows for simple butt coupling, and (c) their modulation capabilities that extend into the gigahertz range [Che88]. Use of both edge-emitting lasers (EELs) and surface-emitting lasers (SELs) has been considered in laser array packaging activities. Whereas EEL manufacturing is a mature technology, recent research in SELs has shown that they are also feasible for optical interconnect applications [Hah95, Sch95]. The advantages of SELs are (1) their ability to be tested at wafer level, which lowers the device cost, and (2) their high coupling efficiency to fiber, which is a result of a single-mode Gaussian output profile.

So far, attention has been focused on sources for one-dimensional arrays for parallel interconnection. However, surface-active devices such as the SELs and SEEDs discussed above allow one to implement two-dimensional optical data links. Furthermore, these devices have been used in photonic switching experiments [McC90], where a three-dimensional optical interconnection offers the potential for a large number of interconnections.

9.2.2 Optical Receiver Arrays

InGaAs/InP PiN photodiodes are used extensively in hybrid optical receivers [Mun94] because of their ease of fabrication, high reliability, and biasing compatibility with electronics. These devices can be produced with a low capacitance, bias voltage, and dark current [Yan93]. Furthermore, it is important to have very low crosstalk and variations in performance across the array. Also, because of the large target, which is on the order of 40–60 μm in diameter, the fiber alignment is relaxed compared to the transmitter packages.

In the following sections, we will discuss packaging issues and the technology available for transmitter and receiver array packaging. The optical fiber arrays,

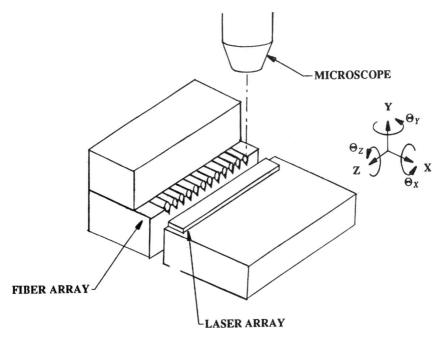

FIGURE 9.1. Degrees of freedom for alignment between fiber and laser arrays.

which are used for taking optical signals out or bringing them into the packages, will be discussed in Section 9.4.

9.2.3 Device Alignment

Active Alignment

For most optoelectronic products manufactured today, coupling the light into or out of a fiber/waveguide is accomplished by active alignment. To achieve a good alignment, one has to work with the six degrees of freedom (d.o.f.): X, Y, Z translations and θ_x, θ_y, θ_z rotations (see Figure 9.1). The fiber array is usually held in a silicon fiber carrier, which is fabricated lithographically, to match the active device spacing.

The laser array is mounted on a silicon or ceramic submount using solder. By fabricating the mounts properly, the θ_z d.o.f. can be eliminated. Traditionally, a rough alignment of approximately ± 3–4 μm is achieved by an image detection scheme using a microscope and a five-axis micropositioner. This positioning minimizes the X and the θ_y d.o.f. The final alignment of ± 1 μm is obtained [Kat92, Hal94] by scanning the SM fiber array on the YZ plane while monitoring the coupled light output from the outermost channels. Finally, the fiber array is mechanically fixed in place using solder. Figure 9.2 [Tak94] shows an eight-channel (200 Mb/s/ch, 100 m long) hermetic laser array module with single-mode fibers which has been assembled using active alignment technique.

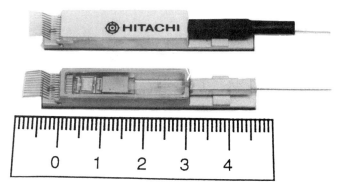

FIGURE 9.2. An eight-channel hermetic laser array module. Courtesy of Hitachi Ltd., reprinted with permission from [Tak94], copyright 1994 IEEE.

Passive Alignment

The active method of alignment described above can be tedious, time-consuming, and expensive. As a result, researchers are actively pursuing passive alignment techniques to reduce the alignment time and cost. Because of the passive nature of the alignment technique, it is important that the piece parts that are aligned be fabricated with the utmost precision. Furthermore, in some cases, additional process steps will be required. Micromachining, which comprises methods used in semiconductor technology such as photolithography, deposition, etching, and so on, is extensively used for achieving the needed level of precision.

Silicon as a "Building Block"

Single-crystal silicon is an excellent mechanical material [Pet82] and is widely used in fabrication of miniature devices and components. Because of its well-established use in the electronics industry, good thermal conductivity, and precision micromachining capabilities, it is an excellent building block available to optoelectronic packaging engineers. The common etch features include V-grooves and pyramidal cavities in $\langle 100 \rangle$-oriented silicon and vertical-walled channels in $\langle 110 \rangle$-oriented silicon. Obviously, the V-grooves are used for locating the fibers and alignment pins, and the cavities for locating the ball lenses. As can be seen in Figure 9.3, the location of fiber with respect to the surface of the silicon is dependent on the quality of silicon, lithography, etching process, fiber diameter, and the core eccentricity. For single-mode applications, it is absolutely necessary to have a well-established anisotropic etching process and fibers with combined diameter and eccentricity tolerance of $< \pm 1 \, \mu m$. The technology for machining $\langle 100 \rangle$- and $\langle 110 \rangle$-oriented silicon by chemical etching is widely published in the literature [Bas78, Bea78, Ken92]. There are numerous chemical etchants available for machining silicon. Two of them, namely EDP (ethylene diamine pyrocatechol) and water, and KOH and water, are quite commonly used due to their versatility. The reported etch ratios [Ken92] are

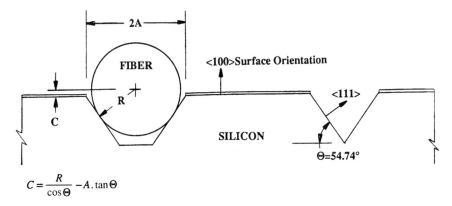

$$C = \frac{R}{\cos\Theta} - A.\tan\Theta$$

FIGURE 9.3. Fiber position in V-groove etched in $\langle 100 \rangle$ silicon.

about $200:100:5$ for the $(100):(110):(111)$ planes in EDP solutions and $200:400:1$ in KOH:water mixtures for the corresponding planes. It is necessary to note that the etch rates slow down considerably as the Si becomes more than $3 \times 10^{19} \text{ cm}^{-3}$ p-type with boron doping.

Epoxy is usually used for the attachment of fibers and ball lenses to the silicon V-grooves and pyramidal cavities, respectively. For hermetic sealing, solder can be used for attaching fibers. This will require additional processing steps of depositing solder-wettable metal on both silicon and fiber surfaces [Bub89]. An alternate attachment technique is the "ALO bonding" [Cou93] under development at Lucent Technologies, which forms solid-state bonds directly between the oxide components (fibers and ball lenses) and aluminum thin-film-coated silicon substrates.

IBM was the first to demonstrate the use of silicon substrate for laser array source packages [Cro77]. The package design included a GaAs laser array bonded to a silicon substrate which also contained V-grooves for an array of multimode fibers. Using a cylindrical lens (a multimode fiber stub) placed between the laser and fiber arrays in a V-groove perpendicular to the fiber array, a coupling efficiency of greater than 50% was realized. Several examples involving the use of etched silicon substrate are described in the next several sections.

Solder Self-Alignment

IBM first introduced the C4 [Mil69, Gol69] (controlled collapse chip connection) technology about two decades ago, which involves fabricating a fusible bump on the IC pad and placing the chip in an inverted position (flipped) on matching bonding pads of the substrate and reflowing the solder. Recently, this technology has been applied to optoelectronic components for achieving both electrical interconnection and optical alignment [Bas93b, LeB94]. There are several advantages in using flip-chip assembly. First, the initial assembly can be performed passively and without the use of a high-accuracy pick-and-place machine. The fine alignment is obtained during the solder reflow process, during which time the surface tension forces pull the chip into the required position. Alignment on the order of 1 μm can be achieved by lithographically controlling the location of pads, controlled formation of solder bumps, and reflowing the solder in a controlled fluxless atmosphere. The presence of a fluxless atmosphere is important in achieving a clean optical surface devoid of flux contamination. Second, large numbers of electrical interconnections can be achieved in a single assembly operation, thereby reducing the assembly cost. Third, the flip-chip technology allows higher switching speeds compared to other types of electrical interconnects such as wirebonding and tape-automated bonding. Fourth, solder bumps provide a good thermal path for transferring heat from the active devices—especially lasers—where heat removal is very critical to their operation. Table 9.1 contains a comprehensive list of solder self-alignment process applications. One of the applications is briefly described below.

Figure 9.4 shows a one-dimensional optical data link (1D-ODL), a parallel interconnection technology [NBH93] developed at AT&T using a strictly passive assembly scheme, which could potentially be used in high-performance switching and computing applications. The transmitter module couples light from a 18-wide edge-emitting laser array into a 18-wide multimode fiber connector located on 250-μm centers. The chip is flip-chip bonded with the epi-side toward the silicon submount using Au–Sn eutectic solder. The silicon submount has been etched on the sides to receive the alignment pins of the fiber connector. A residue-free active atmosphere (gaseous formic acid) solder reflow process [Des92], which does not require any post-cleaning, has been used to keep the laser surface void of contamination. A total of 8 μm of alignment accuracy between the laser array and the multimode fiber array has been reported, inclusive of lithographic inaccuracies, solder self-alignment ($< 3\,\mu$m), laser-active area-to-bonding pad mismatch, and the connectorization scheme.

Figure 9.4b shows the schematic of a receiver module. The packaging scheme and the silicon submount design are identical to the transmitter module except for the incorporation of gold-surfaced turning mirrors in the submount which are used to direct incoming light onto the PiN array. Since the target for PiNs ranges from 40–60 μm in diameter, alignment is usually not a major concern in receiver packaging.

Mechanical "Stops" for Alignment

Etched mechanical features can also be effectively used to align laser arrays with fibers or waveguides. In this case, it will be necessary to have mechanical features etched in both the active array device and the substrate on which the array is mounted. GTE Laboratories introduced this concept [Arm91] by passively aligning a four-element InGaAsP/InP laser array to four single-mode fibers which were mounted in V-grooves of a silicon substrate. Alignment pedestals and standoffs were etched on the silicon substrate such that, when the laser was pushed against the pedestals, each laser was accurately aligned to the corresponding fiber. Figure 9.5 depicts such an arrangement. The laser array was

TABLE 9.1. Examples of Solder Self-Alignment in Array OE Packaging

Assembly	Alignment Obtained or Expected	Wettable Metal	Solder/Size	Reflow Atm.	References
Linear Array					
Coupling waveguide to SM fiber	± 1 μm	Cr–Cu–Au	Pb–Sn eutectic 16 μm high	—	Plessey, UK [WaE90]
LED/PD array on AlN submount	± 3 μm	Cr–Ni	Pb–Sn	—	NEC [Ito91]
Laser array on waferboard [a]	± 1 μm	Ti–Ni–Au	In-solder 9-μm stand off	Flux reflow	GTE [Arm91]
Four-channel transceiver [b]	< 2 μm	Ti–Ni–Au lasers and Cr–Cu–Au for receivers	Pb–Sn eutectic 125-μm diameter	Rosin flux	IBM [Jac92]
LED array on glass substrate	± 2.5 μm	TiW : N–Ni–Au	75-μm diameter	Flux reflow	HP [Iml92]
Laser array on Si submount	< 3 μm	Ti–Pt–Au	Au–Sn < 12 μm high	Fluxless active atmosphere (gaseous formic acid)	AT&T [NBH93]
2-D Array					
Surface-emitting laser array	± 3 μm	Ti–Pt–Au	In 5 ± 0.5 μm high	[c]	AT&T [Jah92]
SEED array	± 1 μm	Ti–Pt–Au	Pb–Sn eutectic 20 μm high	Fluxless active atmosphere (gaseous formic acid)	AT&T [BBB96]

[a] Mechanical alignment with solder bonding.
[b] Solder used to pull chips against mechanical stop.
[c] No reflow performed, planned for future experiments.

fabricated with a lithographically located notch at its edge with respect to the individual lasers. The pedestals and the standoffs were used to obtain alignment in x–y and z directions, respectively. The laser attachment was achieved by depositing a thin layer of solder—less than the height of the standoff—on the substrate, which when reflowed balled up to contact the laser. The solder also provided the electrical contact. This approach yielded an alignment of better than ± 1 μm, with a coupling efficiency of 8%.

In the case of solder self-alignment discussed earlier, the alignment process occurs due to minimization of the total energy of the system. The restoring force,

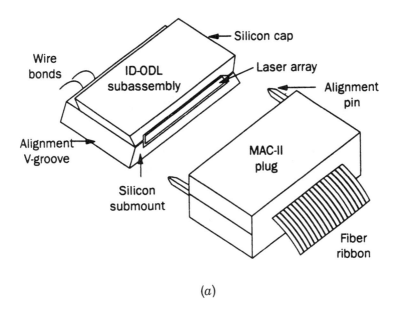

(a)

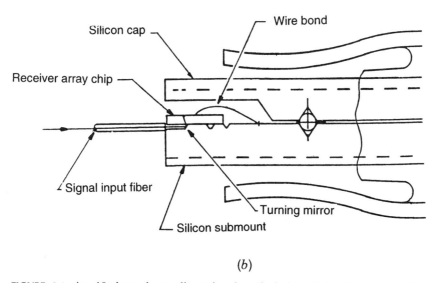

(b)

FIGURE 9.4. An 18-channel one-dimensional optical data link. Courtesy of Lucent Technologies [Nor93], copyright 1993 IEEE.

which is responsible for final alignment, becomes vanishingly small at the final stages of alignment. If alignment can be achieved by locating the chip against a mechanical stop, then the pads can be intentionally offset to provide sufficient restoring force to move the chip to its intended position and hold it in position

(*c*)

FIGURE 9.4 (*Continued*)

during solder freezing. Such a novel example, developed by IBM [Jac92], is shown in Figure 9.6. Here the laser and receiver chips are aligned to waveguides by pulling the chips against lithographically located mechanical stops. An alignment of $< 2\,\mu$m has been reported.

Alignment by Index Method

Another approach based on the registration principles of photolithography is shown in Figure 9.7 [Coh91]. In this method, fiducial marks are placed both on the laser chip and on the fiber carrier. A transparent alignment plate with matching fiducial marks on the bottom side is used to align both the laser chip and the fiber carrier. The positional adjustments in the horizontal plane are made while viewing the fiducial marks through the alignment plate with the aid of a microscope. The relative vertical alignment was achieved by bringing into contact the laser array and the fiber carrier to the bottom surface of the alignment plate. The components are locked in place using solder. Using a AlGaAs ridge SWQ-GRINSCH structure (single quantum well, graded-index, separate carrier confinement, heterostructure) laser array, coupling efficiencies of 8% and 38% for SM and MM fibers, respectively, have been reported.

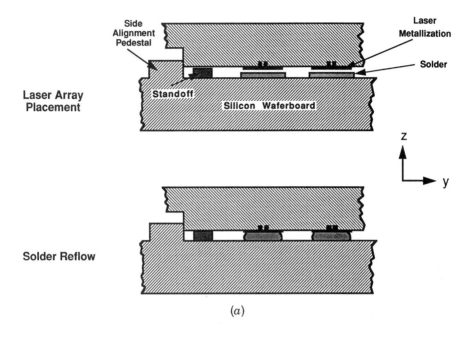

(a)

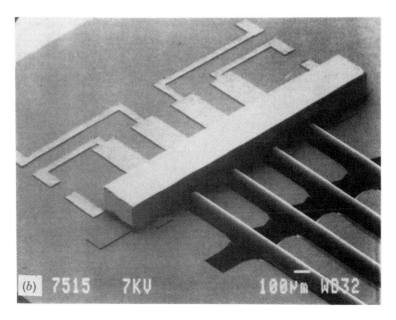

FIGURE 9.5. Passive alignment of laser array using mechanical stops. Courtesy of GTE Labs.

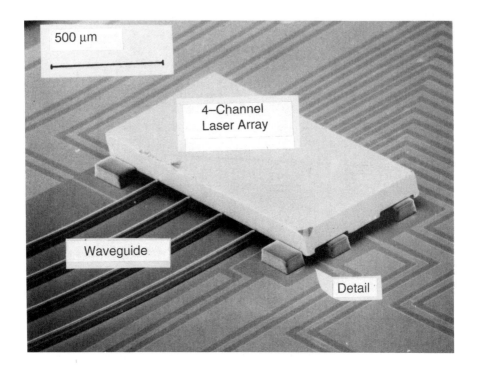

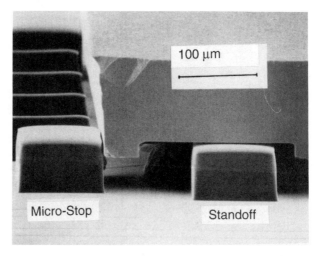

FIGURE 9.6. Flip-chip solder bump attachment for a four-channel laser array. Courtesy of IBM, reprinted with permission from [Jac92], copyright 1992 IEEE.

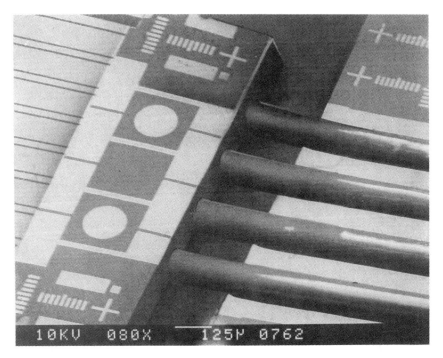

FIGURE 9.7. Laser array alignment using index method. Courtesy of IBM, reprinted with permission from [Coh91], copyright 1991 IEEE.

Figure 9.8 shows another connectorized high-density 32-channel, 16-Gb/s optical data link developed by Optoelectronic Technology Consortium (OETC) [Won95] that uses vertical cavity surface-emitting lasers (VCSELs). The 500-Mb/s/channel NRZ optical data link has been successfully fabricated using multichip module and silicon micromachining technology. The alignment marks on the VCSEL array chip and the base plate are used to position the chip on the base plate. The laser chip is bonded to the silicon baseplate using conducting epoxy. A fiber array block (FAB)—a silicon piece with 32 V-grooves—that carries fiber stubs is epoxied on top of the laser array for transmitting light from VCSELs to a MACII 32-fiber connector. As illustrated in Figure 9.8, light from the laser is coupled into fibers in the FAB using mirrors on the ends of the fibers. The mirrors are formed by angle-polishing (45°) the fiber ends and depositing a thick layer of gold film.

9.3 TWO-DIMENSIONAL SURFACE-ACTIVE DEVICES

For movement of information over short distances, three-dimensional free-space optical interconnection offers the potential for a large number of interconnec-

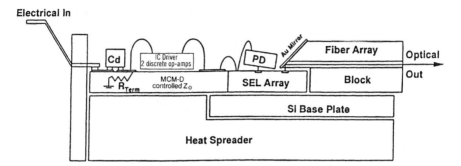

FIGURE 9.8. Schematic of a connectorized 32-channel optical data link using vertical-cavity surface-emitting lasers. Courtesy of Lucent Technologies, reprinted with permission from [Won95], copyright 1995 IEEE.

tions. This provides system designers another avenue for designing large VLSI systems and photonic switching systems [McC90]. These systems require surface-active devices such as SELs and bistable elements like SEEDs. In addition to achieving alignment and high-speed electrical interconnection, uniformity of temperature over the active area of the chip is important in packaging these devices. For example, the optical frequency of maximum contrast in SEED devices is a function of device temperature. Furthermore, because the devices are surface-active, it is advisable to use the surface as a reference while designing the package. This will be helpful subsequently during system assembly where all optics have to be referenced to this surface.

9.3.1 Device Packaging

Figure 9.9 shows the hybrid integration of a SEED device package [BBB96, BaW92] with SEED chips mounted directly onto a quartz substrate containing the microlens array. The substrate serves two purposes: First, it serves as a window for the active devices, and second, it is an optical element of the system. The design uses a flip-chip solder self-alignment assembly to provide both optical alignment and electrical interconnection in one assembly process step. By lithographically placing the I/O pads on the microlens substrate corresponding to the SEED I/O footprint and using a Ti:Pt:Au metallization scheme, an alignment accuracy of $< 1 \mu$m has been achieved. The solder reflow is performed in a reducing atmosphere—gaseous formic acid—without the use of flux, to avoid contaminants on the surface of the SEED chip. The distance between the microlens array and the chip has been controlled to a very high degree of accuracy by depositing Pb–Sn eutectic solder bumps using flash evaporation. In addition to the advantages mentioned in Section 9.2, just like in flip-chip assembly of electronic chips, there is a potential for filling the gap between the chip and the substrate with index-matching epoxy to improve the fatigue life of the joints. A metal tub (heatsink) that encloses the chip is attached using a low-

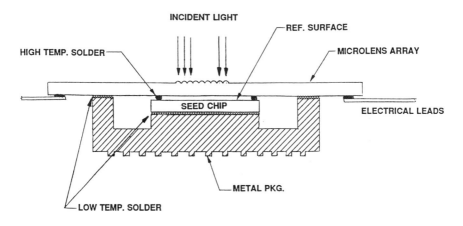

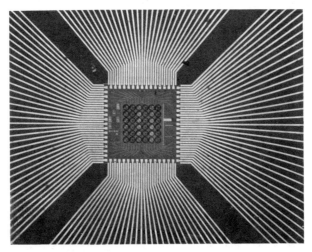

FIGURE 9.9. Two-dimensional surface-active device package. From [BBB96], copyright 1996 IEEE.

temperature indium solder to the substrate to achieve good uniformity of temperature in the devices. It can also potentially be used for hermetic sealing. This type of design lends itself to batch assembly processing, where all the assembly and reflow can be performed on the substrate before dicing it into individual packages. Furthermore, other microlens components can be stacked on top of the lens window (see Section 9.4.2) during system integration.

Another type of package, shown in Figure 9.10 [Olb93], is the monolithic integration of SELs, drivers, and microoptic lenslet arrays. The use of high-speed optical interconnects is one potential application for an array of these smart pixels. Since these devices have strict operational requirements, it is necessary

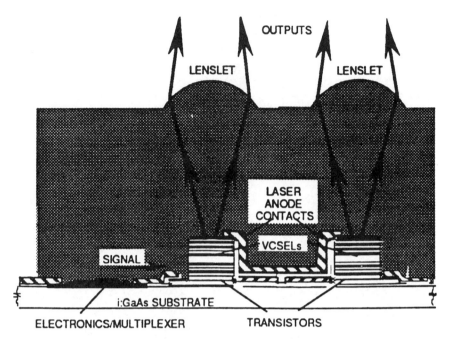

FIGURE 9.10. Monolithic integration of surface-emitting lasers, drivers, and microlenses. Courtesy of Photonic Research Inc., reprinted with permission from [Olb93], copyright 1993 SPIE.

that the packaging technology be developed along with the development of the device technology.

9.4 PASSIVE DEVICES

9.4.1 Fiber Arrays

It is clear from previous discussions and examples that fiber arrays are essential for transmitting optical signals to or from devices. The fibers have to be spaced precisely to match the location of active device arrays. The variation in pitch of the fibers can range from < 1 μm for single-mode fibers to several microns for multimode fibers. In the case of free-space optics, where a two-dimensional fiber bundle is used for inputting pulsed optical signals onto the interconnection fabric using single-mode fibers, the positional tolerance can range from 2 to 5 μm depending on the system architecture and application. The location of light beams emerging from the fibers depends on two factors: (1) location of the fiber and (2) concentricity of the fiber core. Along with the location, it is also important that all beams emerging from the fibers propagate in the same direction collinear to the optic axis of the system. Such requirements translate into fabrication of precision mechanical elements to support and hold the fibers at precise locations.

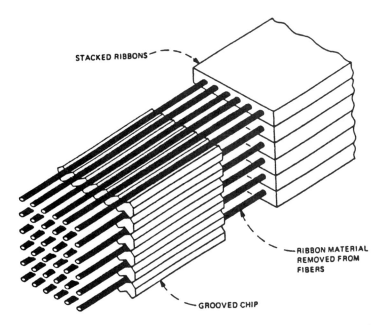

STACKED RIBBONS

RIBBON MATERIAL
REMOVED FROM
FIBERS

GROOVED CHIP

FIGURE 9.11. Two-dimensional fiber array using grooved silicon chips. Reprinted with permission from [Sch78], copyright 1978 AT&T. All rights reserved.

Linear Arrays

The use of single-crystal silicon for aligning active devices has been discussed earlier in Section 9.2. Silicon can also be used for linear fiber arrays to position fibers to accuracies of $< 1 \mu$m. These arrays are also commercially available as MACII[TM] multifiber connectors.[1] Other commercially available linear single-mode fiber arrays are fabricated by using precision molding of silica-filled plastic ferrules[2] and machined V-grooves in ceramic substrates.[3] For building prototypes, the arrays fabricated using ceramic substrates hold an advantage over other substrates because of the machining versatility that allows the creation of a wide range of arrays without incurring extra setup costs.

Two-Dimensional Arrays

Figure 9.11 shows preferentially etched grooves in silicon that have historically been used for splicing arrays of fiber [Sch78, Mil78]. To form a two-dimensional array, it is necessary to stack these silicon chips, which in turn requires that the silicon wafers have two-sided lithography and be etched from both sides. Although high-dimensional accuracy between grooves can be realized, stacking the chips will require tolerances on the wafer thickness to be less than 1μm.

[1] MAC[TM] connectors are products of Berg Electronics, St. Louis, MO, USA.
[2] Manufactured by US Conec Ltd., Hickory, NC, USA.
[3] Manufactured by NGK Insulators Ltd., Japan.

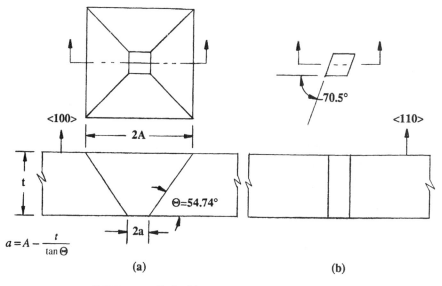

FIGURE 9.12. Etched holes in silicon for fiber array.

Maintaining this tight tolerance on wafer thickness is a very difficult task to achieve.

An alternate method of building fiber bundles is by inserting fibers through a substrate with precisely etched holes [BaW92, Bas92b, Bas93, Pro94]. Obviously, the location of fibers depends on (1) the location of the hole in the substrate and (2) the clearance between the fiber and the hole. Figure 9.12a shows a $\langle 100 \rangle$-oriented silicon substrate with etched holes. The sides of the holes are tapered because they are defined by the $\langle 111 \rangle$ crystal planes. The sloping sides limit the spacing between the holes and thus the fibers. As an example, for a 0.51-mm

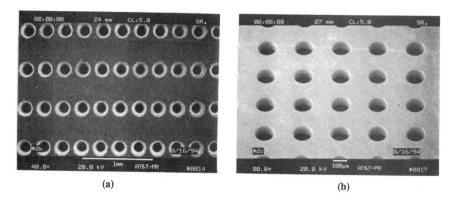

FIGURE 9.13. Etched holes in (a) Fotoform® and (b) brass substrates.

TABLE 9.2. Spacer Material and Fabrication Tolerance

Material	Hole Spacing Tolerance (μm)	Hole Size Tolerance (μm)	Application
Fotoform® glass	±6	±5	MM fiber
Brass	±3	±3	SM/MM fibers
Silicon	±0.5	±2	SM/MM fibers

(20-mil)-thick wafer, the minimum hole separation that can be obtained is $\sim 845\,\mu$m. Etching holes from both sides of the wafer reduces the minimum separation to half the above value. However, this method requires precise alignment of masks for the two-sided lithography.

Meanwhile, fabricating substrates with holes from silicon with $\langle 110 \rangle$ orientation will feature parallelogram-shaped holes with vertical walls [JPL87], as shown in Figure 9.12b.

Other types of substrates that are suitable for etching holes are photosensitive glass materials such as Fotoform® made by Corning, Inc. and thin brass stock as shown in Figure 9.13. It can be seen that the brass gives a smoother etched wall surface compared to Fotoform material. Since the thickness of brass substrates is on the order of 100 μm, a backing substrate such as Fotoform glass is needed to strengthen the structure.

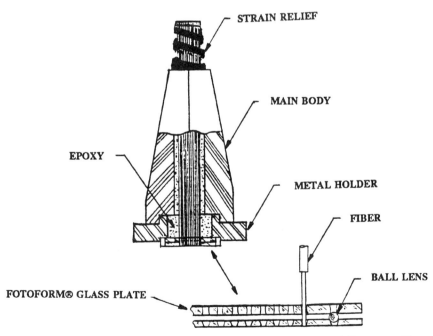

FIGURE 9.14. Schematic showing the construction of a two-dimensional fiber array. From [Bas92], copyright 1992 ASME.

Table 9.2 compares the tolerances on the locations and diameter of holes in various substrates. Depending on the system requirements and the type of fiber being used—single-mode or multimode—one can choose the appropriate substrate.

Fiber Array Design

Figure 9.14 shows the concept for constructing a fiber array. Two substrates (in this case Fotoform® glass) are aligned, with one stacked above the other using precision sapphire spheres, and epoxy bonded. Stacking the substrates serves two purposes. First, tapering the back substrate provides a lead-in for ease of insertion of fibers, and second, it maintains the fibers in vertical position after insertion. The substrate shown in Figure 9.13a is used in this design. The fiber bundle is fabricated by inserting fibers through the holes in the stack, epoxying the fibers in place, and polishing the ends.

It is obvious that forming large 2-D arrays by manually inserting fibers without causing fiber damage can be tedious at best. Hence, an assembly apparatus as shown in Figure 9.15 is needed for simultaneously inserting 12–18 fibers into the substrate.

One of the main concerns in any fiber-integrated component is fiber breakage during fiber handling. Another concern when fabricating large fiber bundles is

FIBER RIBBON

FIBER INSERTION TOOL

X-Y STAGE

FIGURE 9.15. Assembly apparatus for inserting multiple fibers into a substrate. From [Bas92], copyright 1992 ASME.

the development of cracks during polishing. These problems can be partially overcome by using a modular approach—a manufacturable one—as shown in Figure 9.16 [Bas96]. Here, linear fiber arrays with alignment pins precisely located with respect to the fibers are placed next to each other. The alignment pins are used for plugging the individual arrays into a cage to form a 2-D fiber array. Even though the distance between arrays is controlled to high accuracies using either V-grooved etched silicon or machined ceramic spacers, this distance is relatively large since it is dictated by the thickness of individual arrays. Figure 9.17 shows the alignment that can be realized with such a modular approach.

Examples of different approaches used in fabricating prototypes of fiber arrays are tabulated in Table 9.3.

9.4.2 Microlens Arrays

Microlens technology is being increasingly used in a wide range of optical system applications. Some of the applications include microlens arrays attached to fiber arrays for inputting optical signals onto free-space switching fabric, enhancing efficiencies of detector arrays, imaging arrays for facsimiles, image transfer

FIGURE 9.16. Modular 2-D fiber array using linear arrays. From [Bas95b], copyright 1995 OSA.

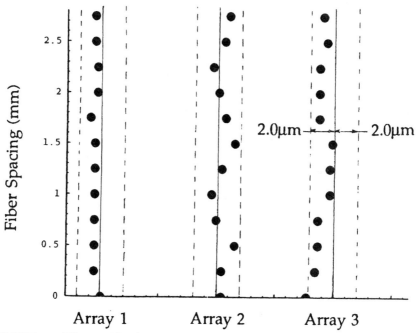

FIGURE 9.17. Fiber position in a modular 2-D fiber array. From [Bas95b], copyright 1995 OSA.

systems, liquid crystal display throughput enhancement, and so on. One of the key elements in stacked planar optics [IgO82] is precisely aligned 2-D microlens arrays. These lenses are fabricated to photolithographic tolerances on planar wafers [Mer92]. In this section we will discuss some of the possible ways of passively aligning microlens arrays.

Alignment of Microlens to Fiber Arrays

Since the Fotoform® and brass substrates shown in Figure 9.13 are fabricated using photolithography steps, it is conceivable to incorporate mechanical alignment features on these substrates. Figure 9.18 is a photograph of a microlens array attached to a linear MM fiber array (MACII) using a photo-chemically etched brass holder. Microlens-to-fiber alignment has been obtained by using the alignment pins as guides for the brass holder.

Figure 9.19a shows the concept of attaching a microlens array onto a 2-D fiber array. Alignment microlenses, which are fabricated lithographically along with the microlens array, are used to attach the microlens array to a brass substrate holder. The holder in turn is aligned and attached to the fiber array using precision glass ball lenses. These ball lenses can be obtained with a diameter tolerance held to ±1 μm. Figure 9.19b shows a photograph of a microlens array attached to a 2-D fiber array.

TABLE 9.3. Fiber Array

Size/ Fiber	Fiber Spacing (μm)	Alignment Accuracy (μm)	Construction	Reference
		Linear Array		
1 × 18/SM and MM	x-250	1.0 (expected)	V-grooved silicon (MAC-II)	ATT/Berg Electronics, USA
1 × 18/SM and MM	x-250	1.0 (expected)	Molded plastic	NTT, Japan
1× N/SM	x-250	< 1.0	Machined ceramic	NGK-LOCKE, USA
1 × 80	x-250	1.0 (expected)	Molded plastic	NTT, 1993
		2-D Array		
12 × 12/MM	x-228.6, y-?	2.5	Stacking V-grooved silicon	ATT [Sch78]
37/SM-MM combination	285	12.6	Feeding individual fibers through 150 μm steel substrate with holes	COMSAT [KoM84]
8 × 16/MM	x-262, y-524	10.0	Inserting multifibers through etched holes in Fotoform® substrates	ATT [BaW92]
12 × 12/MM	x-250, y-250	< 5.0	Inserting multifibers through etched holes in brass substrates	ATT[Bas92a]
8 × 8/SM	x-250, y-250	5.0	Stacking μ-ferrules	NTT [Koy93]
4 × 4/SM	x-200, y-200	< 10.0	Inserting fibers through laser-drilled holes in polyimide sheets	British Aerospace [Pro94]
4 × 4/MM	x-800, y-800	< 20.0	Inserting fibers through holes etched in silicon substrates	British Aerospace [Pro94]
4 × 8/SM	x-250, y-250	1.5	Individual fiber-active alignment using template	ATT [Sas94]
5 × 12/SM	x-250, y-3500	< 3.0	Modular—plugging-in linear arrays into a cage	ATT [Bas95b]
4 × 72/SM	x-250, y-4000	< ?	Modular—plugging-in linear arrays into a cage	ATT [Bas95a]

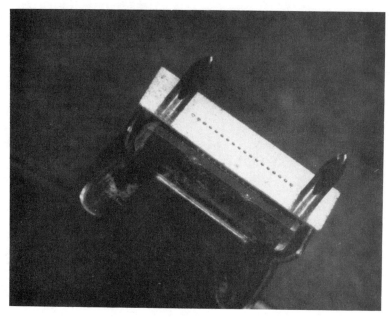

FIGURE 9.18. Microlens array attached to a linear array. From [Bas95b], copyright 1995 OSA.

Stacking of Microlens Arrays

Figure 9.20 shows different methods of aligning microlens arrays [Bas94]. In Figure 9.20a, silicon substrates with etched cavities and holes, along with precision ball lenses, have been used for aligning two microlens arrays. The distance between microlenses, which have lower spacing tolerance compared to the lateral alignment tolerance, can be easily varied by using ball lenses with different diameters.

Solder self-alignment can also be used to align microlens arrays, as shown in Figure 9.20b. Since a precise amount of solder has to be deposited uniformly for forming solder bumps, the spacing is usually limited to $100\,\mu$m. Also, the precision of lateral alignment is dictated by the two-sided lithography (needed for depositing the metal and solder on the back side of one of the arrays) and the solder self-alignment process.

A somewhat simpler structure is depicted in Figure 9.20c. As shown, three sets of three alignment microlenses are formed in conjunction with optical microlens arrays. Each alignment microlens set forms a tripod within which can nest a ball lens spacer. By using three ball lenses on one of the arrays—mounted, for example, using epoxy or ALO bonding technique described previously in Section 9.2—and placing the second array in a flipped position, an alignment of $< 1\,\mu$m between the two microlens arrays can be easily achieved.

It is possible for one to conceive different combinations of alignment schemes using the above methods.

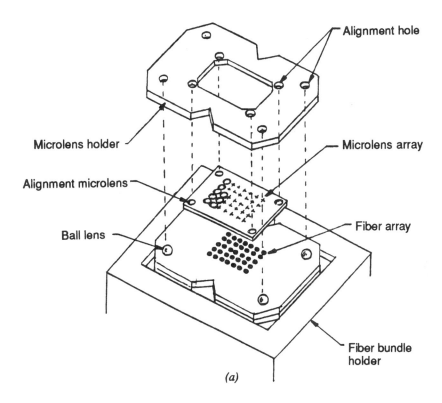

Alignment hole

Microlens holder

Alignment microlens

Ball lens

Microlens array

Fiber array

Fiber bundle holder

(a)

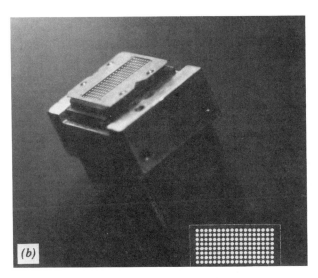

(b)

FIGURE 9.19. Two-dimensional microlens array attached to a fiber array. From [Bas95b], copyright 1995 OSA.

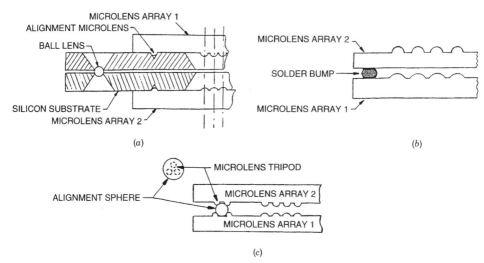

FIGURE 9.20. Various concepts showing stacking of microlens arrays. From [Bas92a], copyright ASME, and from [Bas94].

9.5 SUMMARY

Future systems such as a BISDN, advanced switching, and computers require higher-performance and lower-cost packaging technologies. Parallel optical interconnection technology will be a necessity for such systems. In this chapter, we have discussed the current status of array technology, which is still being developed in various laboratories. Deployment of this technology will mainly depend on the packaging cost and introduction of new lasers such as expanded mode lasers, which have relaxed fiber-to-laser beam alignment.

CHAPTER TEN

Hybrid Technology for Optoelectronic Packaging

PAUL HAUGSJAA

When it becomes necessary to reduce the cost and size of an optoelectronic system or to increase its functionality or reliability, engineers usually look toward increased integration to meet these needs [Wad94]. Integration is usually achieved by combining devices in a common assembly that can be tested and packaged as a unit to provide an enhanced level of functionality. Monolithic integration utilizes a single device substrate on which all the devices in the assembly are fabricated. Due to the varied materials requirements for complex optoelectronic devices, a monolithic approach necessitates compromises in the selection of materials parameters, device design, and the processes used to fabricate the integrated assembly. In some cases this can be limiting to performance and yield, since only a single material type may be used and compromises may be extensive. Monolithic integration does, however, offer some distinct advantages for potential cost reduction, manufacturing yield enhancement, and the provision of a structured design environment [YKN95, LCG95]. These advantages are well known in, for example, the case of the electronic integrated circuit. Hybrid integration, on the other hand, brings several individual and potentially dissimilar elements together. It is often the case that hybrid integration can utilize monolithic optoelectronic integrated circuits such as array receivers [HSY94, UYH94, PPR93, BBF95] combined with other components in a hybrid assembly. One special form of hybrid integration uses thin-film optoelectronic devices attached to an integrated circuit fabricated in another materials system [GWD95, Jok94, SCA91]. Hybrid integration thus allows diverse technologies to be combined, despite each being optimized for a particular function. This form of integration thus offers some clear advantages in the case of optoelectronic subsystems where vastly varied device functionality must be provided, albeit with lower yields, probably, than could be achieved with monolithic integration. This chapter will discuss the packaging of optoelectronic subsystems that have been implemented via hybrid integration. The packaging of hybrid integrated elements is quite similar in many respects to packaging of hybrid electronic integrated circuits or multichip modules (MCMs), with some critical variations. Thus, the very natural migration of optoelectronic integration into the realm of MCMs will also be discussed in this chapter.

10.1 PLATFORMS FOR INTEGRATION

Selection of the substrate for hybrid optoelectronic integration must be carefully considered since it must fulfill a variety of functions. Hybrid integration usually combines several component technologies on a single mounting platform. Thus, the platform must be compatible with all of the materials and interface requirements of the individual component technologies. Table 10.1 lists the elements of a hybrid optoelectronic integrated circuit (OEIC) with their respective functions.

Since a packaged component is the desired end result, it makes sense to choose a mounting platform that can ease the task of packaging and solve as many problems as possible for the packaging engineer. As described in Chapter 8, chemical compatibility, matched coefficients of thermal expansion, appropriate thermal conductivity, and the selection of suitable temperature sequences for the assembly operation are important for the entire subassembly. Each of the individual components and the platform substrate itself must be considered for these concerns. Issues involving chip attachment, wire bonding, flip-chip solder bonding, passive and active optoelectronic alignment, electronic chip bonding, control of reflections, internal package atmosphere, minimization of undesired electronic parasitic impedances, and so on, must be considered in like manner. Since the resulting packaged component will undoubtedly be capable of rather broad functionalities, it will be useful to include a means of connection to other optoelectronic systems by including optical connectors in the packaging concept and providing a means of electronic interconnection that is entirely consistent with interconnections to other hybrid subassemblies or MCMs.

Ceramic substrates can represent a good choice for use as a packaging platform, as discussed in Section 10.4. Certainly the technology for patterning metal and fabricating appropriately shaped ceramic platforms has been well developed for use in the hybrid circuit and MCM industries. Additionally, technology for the machining of precision grooves for the alignment of optical fibers and other package elements into ceramic, glass, or other substrates makes them acceptable for platforms with micrometer alignment requirements.

Often, silicon substrates are utilized in the fabrication of hybrid integrated circuits and can result in densely interconnected, yet possibly quite large, wafer-

TABLE 10.1. The Elements That May Make Up a Packaged Hybrid Optoelectronic Integrated Circuit (OEIC) Along with Their Respective Functionalities

Hybrid OEIC Element	Function
Mounting platform or substrate	Allows hybrid integration of all elements
Electronic integrated circuits	Provides complex electronic functionality
Discrete electronic components: transistors, capacitors, resistors, etc.	Supplies specific electronic circuit elements
Optoelectronic devices: lasers, detectors, acoustooptic filters, etc.	Provides for photonic signal emission, detection, and processing
Optoelectronic integrated circuits	Allows complex optoelectronic functionality
Guided-wave optical devices: optical couplers, switches,	Gives optical routing and optical signal modulation
Optical elements: lenses, filters, mirrors, etc.	Provides optical functionality
Optical guiding elements: optical waveguide, optical fiber, etc.	Routes optical information into and around hybrid OEIC
First-level package	Provides physical protection and heat spreading capabilities for the hybrid OEIC
Electronic input/output feed throughs	Routes electronic information, power, and ground connections in and out of package
Optical input/output feed throughs	Allows optical signals to pass through package walls with stress relief, and hermetic sealing yet without excessive optical loss or polarization shifts
Environmental sealing	Maintains integrity of package atmosphere (hermeticity)
Heat sink	Removes any heat generated in the assembly
Alignment elements: notches, pedestals, standoffs, V-grooves, etch pits, precision spheres	Assist in the precision alignment of optical and optoelectronic elements within package

scale integrated assemblies. Formation of optical waveguides on the silicon substrate may be accomplished using glass [HBK89, TYK86] or polymeric depositions [FLH95]. As we shall see, most applications to date utilize substrates that are only a small portion of the size of a semiconductor wafer, but the benefits of treating integration on the wafer scale bring significant manufacturing advantages. Silicon substrates may contain built-in electronic components such as termination resistors and decoupling capacitors, as well as transistors and even waveguides. An added advantage offered by silicon substrates is the substantial capability for micromachining grooves, precision microstops, and a variety of optical elements [LLP94]. Section 10.4 details use of silicon substrates for alignment purposes. Techniques used include orientation-dependent etching of the single-crystal substrate to reveal precisely angled crystal planes useful for alignment of optical fibers, spherical lenses, optical waveguides, and so on [MYY94, JCN94]. The micromachining capability may also be applied to formation of pyramidal fiducial cavities that, when combined with small alignment spheres, can provide registration between several substrates [KHH94, DBW95]. Alternatively, optical fiber glass rods can provide alignment between two or more substrates with precision-etched V-grooves [CSC94, KGK96]. Good use of silicon micromachining capabilities has been made in passive alignment studies involving laser arrays and single-mode fiber arrays [ATJ91] and also in studies including expanded-mode semiconductor lasers to achieve low-cost, high-efficiency packaging [LSR94]. Additionally, silicon substrates have been used in the alignment of arrays of semiconductor optical amplifiers to single-mode optical fiber arrays [LBN95]. Recently, micromachined optical components have been fabricated on silicon substrates with built-in passive alignment capabilities [LLW95].

Plastic substrates have been applied to the fabrication of optical connectors for both single- and multimode optical fiber [NYA91]. These quite precise and stable plastic alignment subassemblies with precision pin registration are molded with new, dimensionally stable silica-filled polymer materials. Plastic materials of this sort should thus be considered as a possible integration platform for use in optoelectronic packaging. Advantages could result, particularly in reducing costs and simplifying the fabrication process.

10.2 WAFER-SCALE MANUFACTURING OF OPTOELECTRONIC INTEGRATED CIRCUITS

If sizable substrates are used that allow dense interconnections on the substrate, it is possible to fabricate assemblies with a high degree of complexity and functionality across a large area. In the world of electronics, this sort of electronic integration is often referred to as wafer-scale integration. When applied to optoelectronic integration, wafer-scale manufacturing processes provide for the added functionality of optical waveguides and optoelectronics to be included. In utilizing wafer-scale manufacturing, an added benefit in the mass fabrication of integrated optoelectronic subassemblies results. Often, processes that have been developed to aid in the fabrication of electronic integrated circuits on semiconductor wafers can be applied to the fabrication of optoelectronic integrated circuits assembled on semiconductor wafers or other wafer-like platforms. Wafer handling methods, thin-film processes, characterization techniques, wafer probing methods, optical testing procedures, process control techniques, die separation processes, and binning/marking techniques can all be applied to advantage in fabricating large numbers of components and testing many assemblies produced by hybrid optoelectronic integration to provide good yields, high reliability, and relatively low costs. Even components of relatively low complexity may be fabricated using wafer-scale integration processes for the cost advantages provided [DaB94].

IBM has developed a wafer-scale process for the testing of laser diodes, where lasers may be tested prior to the die separation process. This allows high-speed, low-cost characterization and identification of known good laser die [VBB91]. A photograph of the etched-facet, edge-emitting lasers that were developed to allow on-wafer testing of completed lasers is shown in Figure 10.1.

10.3 FIRST-LEVEL PACKAGING OF OPTOELECTRONIC INTEGRATED ASSEMBLIES

The packaging solutions applied to first-level packaging of hybrid optoelectronic integrated assemblies have evolved, in most cases, directly from the packaging of

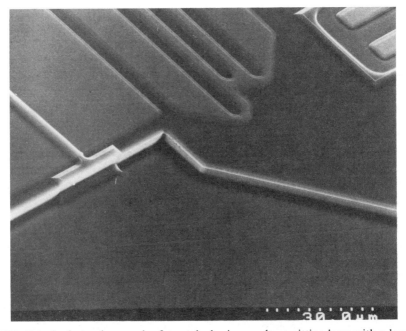

FIGURE 10.1. A photomicrograph of an etched-mirror, edge-emitting laser with a large-area reflecting chute to allow on-wafer testing of fabricated lasers [VBB91]. Reprinted with permission of IBM Zurich Research Laboratories. Copyright 1991 IEEE.

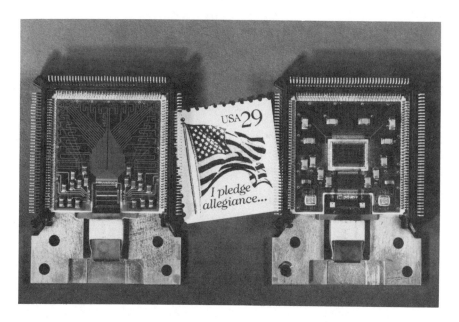

FIGURE 10.2. A photograph of the optoelectronic integrated 32-channel parallel optical interconnect receiver and transmitter hybrid integrated assemblies that have been fabricated by the OETC Consortium [Won95]. Reprinted with permission of Lockheed Martin and AT&T Bell Laboratories. Copyright 1995 IEEE.

integrated circuits (ICs), multichip modules (MCMs), and discrete optoelectronic components [TuR89]. Packaging for discrete optoelectronic components has already been discussed in Chapters 5, 6, and 8. Packaging of integrated optoelectronic assemblies really involves an application of these same principles with the added—perhaps obvious—concern that packaging solutions for the various elements of an integrated assembly must not interact unfavorably. Materials for first-level packages include metal cans with ceramic or glass feedthrough elements, ceramic enclosures, and plastic assemblies such as the one shown in Figure 10.2, developed by the Optoelectronic Technology Consortium (OETC), including Martin-Marietta, AT&T, IBM, and Honeywell with support from ARPA. These hybrid integrated optical interconnect modules use a silicon subassembly to align the optical fibers and turn the optical beams on beveled fiber mirrors to allow coupling to the planar OEIC detectors and to a vertical-cavity surface-emitting laser (VCSEL) array [Won95].

The silicon baseplate provides the alignment to short sections of multimode fiber and pin alignment registration for MAC II^{TM}-type connectors for ribbon optical fiber. A multichip module fabricated on an alumina substrate provides fanout for electrical interconnections and a site for receiver array integrated circuits. The premolded plastic quad flatpack lead frame used for the modules has a 25-mil lead pitch along three sides, for a total of 123 I/Os. Data transfer of more than 250 Mb/s is possible with these modules on each of the 32 optical fibers.

New plastic-based materials are being applied across the industry that offer helium leak-tested levels of sealing approaching levels considered to be hermetic to optical fiber and electronic feedthroughs [JNL95]. Minimal first-level packaging solutions are being developed that eliminate much of the complexity of prior optoelectronic packaging of components. An example of the sort of solutions that are being considered is shown in Figure 10.3. This packageless protection of a laser by BT Laboratories utilizes high-purity silicone gel and passive alignment of an expanded mode laser to achieve a potentially extremely low-cost laser for local loop applications [CPF95]. Multichip modules are also being fabricated with silicone resin encapsulation of bare chips to provide hermetic conditions with little penalty in size [TNM93].

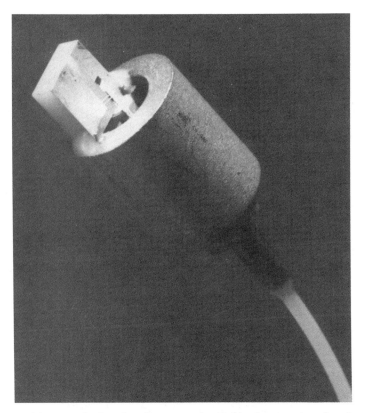

FIGURE 10.3. Photograph of packageless protection and lead connections for a laser using silicone gel coatings and integration platform mounting of a laser by BT Laboratories. Photograph courtesy of BT Laboratories, copyright 1996.

10.4 MERGING OF OPTOELECTRONIC INTEGRATION AND MCM TECHNOLOGY

Multichip modules (MCMs) are finding increasing use in the electronics industry. Multichip modules are often used to provide improved system performance over existing levels of integrated circuit and printed circuit boards. It is necessary with continuously improving electronic integrated circuit technology for engineers to carefully consider the choice of either MCMs or monolithic integration in order to provide a given system element functionality. However, there continues to be a performance advantage in using multichip modules to provide some system benefits (often size and performance) beyond that attainable with any given level of monolithic integration. One advantage provided by multichip modules is a reduction in the number of pin-lead-through elements in an electronic subassembly. Since an MCM needs only a single first-level package, all integrated circuits contained in that package do not have to communicate with the other ICs in the package through the several pins normally required for each electrical interconnection. The fact that package reliability is closely related to the required number of lead-throughs means that, in addition to potential cost and performance advantages, there are definite reliability advantages to be obtained through the reduction in the quantity of packages as well as the number of input/output pins used in a system.

Further reductions in MCM pin count can be obtained by using optical interconnections to provide high-speed data communications between packaged MCMs [Hau95]. In this manner, a 16-bit-wide bus that would normally require approximately 20 pins for operation with a 50-MHz clock could utilize a single optical fiber operating at about 1 Gb/s to transfer the same information. As data rates on electronic interconnections increase, crosstalk and radiated electromagnetic interference problems increase also. Thus, as the amount of data handled by a system increases, it is clear that use of optical interconnections can provide distinct advantages for communications between MCMs. Clearly, this will only happen if the optical interconnections are low in cost and at least as reliable as the electronic interconnections that they replace. A great deal of work is presently underway in the development of optical interconnects that can be applied both to board-to-board interconnections and to communications between multichip modules. Optical waveguide interconnections between multichip modules with optical waveguide structures that can be fabricated in a manner similar to electronic interconnections are being developed for commercial MCM fabrication lines [FWB94]. In addition, polymeric waveguides can be used to fabricate active optoelectronic components such as modulators and switches [FLH95] to further increase the functionality of MCMs.

Typical requirements for digital optical interconnections are given in Table 10.2. These requirements are often related to the rate of transferred data provided by the link. For example, higher-data-rate links can support a higher cost per link, since fewer links will be required for any given aggregate of data to be transferred. It is important to note in this regard that the total system

TABLE 10.2. Typical Performance Requirements for Digital Optical Interconnects

Parameter	Requirement
Distance	Up to 200 meters
Hermeticity	As required to meet reliability required
Failure rate	< 1 failure in 10^7 hours
Data rate	500 Mb/s to 2.5 Gb/s per optical fiber
Error rate	Less than one error in 10^{12} bits transferred prior to error correction
Operating temperature	-10 to $+60°C$ system ambient without thermal electric coolers desirable
Storage temperature	-30 to $+85°C$ for long periods, solderable at $210°C$ for short periods desirable
Cost	$< \$100$ per Gb/s of transferred data
Size (footprint)	< 3 cm^2 per Gb/s of transferred data

requirements (cost, performance, and reliability) may in the end dictate the choice of interconnect technology.

Though the discussion of hybrid integration thus far has concentrated on digital systems, in fact, hybrid integration of RF and microwave optoelectronic systems has been and continues to be an important application for such integration. Analog optical interconnect requirements have additional requirements for linearity and faithful reproduction of signals without addition of intermodulation products, noise, and spectral growth.

The optoelectronic packaging industry is using optical interconnections effectively as a testing ground for new concepts in packaging that can benefit from cost reductions inherent in high-volume applications. Sandia National Laboratories is working on a concept that will allow three-dimensional free-space interconnections between MCMs [CSC94]. Several forms of optical interconnections for board-to-board or intra-MCM applications have been developed that are excellent examples of hybrid wafer-scale packaging of optoelectronic assemblies. MIT Lincoln Laboratory has recently developed a free-space optical interconnect that is a good example of hybrid optical integration [TRW94]. This interconnect uses OEIC receivers, laser arrays, and integrated microlens assemblies integrated on ceramic platforms to provide an interconnect that is capable of 2-Gb/s data transfer over 20 free-space optical channels at a separation of 5 mm. A photograph of this optical interconnect is shown in Figure 10.4.

A consortium of organizations including Hewlett-Packard, Du Pont, AMP, the University of Southern California, and SDL, called the Parallel Optical Link Organization (POLO), is developing an optical interconnect with the partial support of ARPA that utilizes VCSEL arrays to provide a data link utilizing 10 multimode optical fibers with an aggregate data rate of 10 Gb/s [Hah95]. POLO

FIGURE 10.4. A photograph of a free-space optical interconnect developed by MIT Lincoln Laboratory that is an example of hybid optoelectronic integration [Tsa95]. This interconnect includes lens arrays, OEICs, laser arrays, and ceramic integration platforms. Reprinted with permission of MIT Lincoln Laboratory, Massachusetts, copyright 1995 OSA.

has worked on optical interconnects that use multimode optical waveguides fabricated with Du Pont Polyguide™ material beveled to provide turning mirrors for coupling to planar detectors and vertical-cavity surface-emitting lasers. These waveguide sections are coupled to optical fiber ribbon with a clip connector [BMC94]. Figure 10.5 shows a prototype transmitter that uses multimode optical fibers on a pitch of 250 μm operating at 1 Gb/s.

IBM developed another good example of a hybrid integrated optoelectronic circuit assembly. This assembly consisted of a single-channel optical transmitter and receiver in a single package [DDF92]. A photograph of the packaged assembly is shown in Figure 10.6. This transceiver module operated at a data transfer rate of 1.06 Gb/s over single-mode optical fiber with a bit error rate of less than 10^{-14}. It utilized a 1.3-μm laser and a pin detector for data transfer that was compatible with the ANSI Standard X3T9.3 Fibre-Channel Protocol. IBM took care in the selection of the optical fiber connectors for these hybrid assemblies that they should be rather small but also provide eye protection in the case of an open transmitter port by incorporating a shutter assembly into the connector. Though use of tight-tolerance plastic parts was made in this assembly, a large portion of the cost of the component remained in the optical subassembly. Since this transceiver proved to be too expensive to manufacture for the

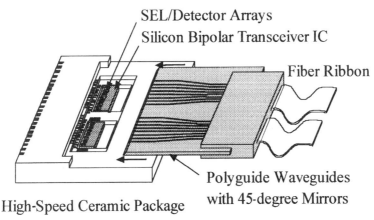

SEL/Detector Arrays

Silicon Bipolar Transceiver IC

Fiber Ribbon

Polyguide Waveguides with 45-degree Mirrors

High-Speed Ceramic Package

FIGURE 10.5. A prototype of the parallel optical interconnect assembly developed by the Parallel Optical Link Organization (POLO) utilizing beveled polymer waveguides to couple to VCSEL emitters and planar detectors [Hah95]. Reprinted with permission of Hewlett-Packard Laboratories, copyright 1995 IEEE.

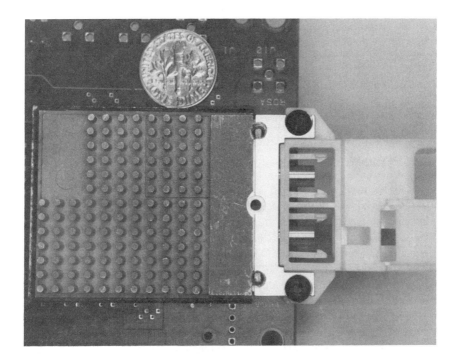

FIGURE 10.6. Photograph of IBM hybrid integrated optoelectronic assembly as an example of a hybrid integrated optical interconnect module. Photograph by P. Haugsjaa, copyright 1996.

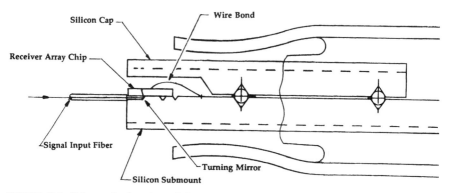

FIGURE 10.7. Schematic drawing of the coupling element in optical receiver module for optical interconnects built by AT&T Bell Laboratories [NBH93]. Reprinted with permission of Lucent Technologies, copyright 1993 IEEE.

market niche it addressed, IBM has pursued other optical interconnect products. It is clear that cost of optical integration can have great effects on the marketability of products. Packaging must be selected that allows extremely low-cost assembly and wide applicability of a product in order to be successful.

AT&T Bell Laboratories (Lucent) has applied ⟨100⟩ silicon substrates as alignment platforms for passive alignment of lasers and detectors to multimode optical fiber via flip-chip solder bumps [NBH93]. This technique uses a silicon cap that has also been etched with V-grooves that match ones on the substrate to provide alignment to the pins of a MAC II™-style connector and thus provide a connectorized optical fiber interconnect to a MCM style package (see also Figure 9.4, Section 9.2). This package is a quad flatpack 84-pin package with a premolded plastic collar around the MCM substrate. The coupling element of the optical receiver module of this interconnect is shown in Figure 10.7. The interconnect provides 18 1-Gb/s channels on a pitch of 250 μm.

Hybrid integration of arrays of analog optoelectronic receivers and transmitters is a rather difficult task considering the dynamic range and linearity requirements of most analog optoelectronic applications. Work on the development of analog transmitters and receiver modules including array transmitters and array receivers is being conducted by a consortium of companies (Hughes, Boeing, GTE, Amoco, United Technologies Photonics, Lincoln Laboratory, and NCCOSC) with the partial support of ARPA under a Technology Reinvestment Program. This consortium is attempting to explore the cost and performance tradeoffs for integration of multiple analog optical interconnect channels on silicon waferboard [HYN95]. A notch process for the laser array has been developed that allows passive alignment of the laser array to a fiber array using micromachined silicon waferboard substrates.

Honeywell and GE have been working on techniques of bringing optical signals onto multichip module substrates to distribute optical signals within the MCM and provide board-level interconnection using polymer optical wave-

FIGURE 10.8. A photograph of the Motorola Optobus^{TM} optical interconnect. Photograph courtesy of Motorola, copyright 1996.

guides. The MCM substrates they are exploring are closely related to those used for the high-density integration (HDI) MCM-D process [LiB95].

Motorola has announced a new optical interconnect product, Optobus^{TM}, that uses a VCSEL array as an optical source with alignment accomplished within a plastic waveguide section. This optical interconnect solution, shown in Figure 10.8, is capable of providing an aggregate data transport of 2.5 Gb/s over 10 optical fibers [SCF95]. While the optoelectronic portion of this product is capable of higher data rates, the technology used in the initial driver and receiver

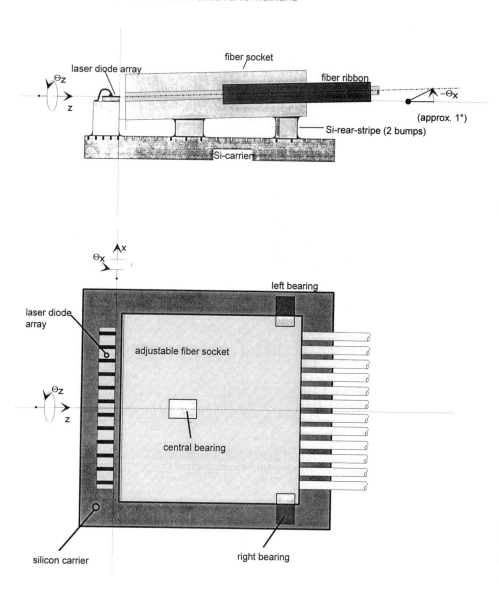

FIGURE 10.9. A schematic drawing of the alignment section of Siemen's laser array transmitter for optical interconnects. The adjustable fiber socket is positioned for optimum alignment by moving the bearing points [KHH94]. Reprinted with permission of Siemens AG, copyright 1994 IEEE.

electronics and the packaging of the module establish limitations to the current data rate that may be upgraded in the future.

Siemens has developed an optical interconnect using lens arrays for coupling to photodiode arrays and low-threshold edge-emitting laser arrays [KHH94]. In

FIGURE 10.10. Micrograph of a 12-channel Siemens optical interconnect receiver, with silicon integration platform and arrays of photodiodes and amplifiers. Photograph courtesy of Siemens AG, copyright 1996.

this work, Siemens has explored a hybrid integration scheme where alignment for multimode optical fiber coupling to lasers is accomplished with optical fibers mounted in a silicon holder with micromachined V-grooves that is fixed at three points: a central movable bearing and two silicon rear bumps of slightly different height than the central bearing. This coupling scheme is shown in Figure 10.9. Movement of the central bearing changes the height adjustment of the optical fiber array. The Siemens optical link, shown in Figure 10.10, is compatible with ECL voltage levels and provides an aggregate data throughput of up to 12 Gb/s through an array of 12 optical fibers on a 250-μm pitch.

The European community's RACE program has developed several versions of fully integrated chip-to-chip and board-to-board optical interconnections. These components utilize multimode optical fiber held in V-grooves aligned to optical and optoelectronic components precisely mounted by Au–Au thermo-compression flip-chip bonding processes at a rather low temperature (300°C)

[DDA94]. In this work, silicon waferboard assembly allows eight optical fibers to be butt-coupled to a 980-nm-wavelength laser array. In the receiver, anisotropically etched and gold-plated end facets of V-grooves serve as turning mirrors for coupling of optical fibers to 80-μm-diameter PIN detectors. The silicon waferboard in this case was mounted on a standard thick-film substrate for interfacing to a system printed circuit board in a manner commonly used for multichip modules. These modules were used in experimenting with coupling to a flexible optical backplane [DVV92] at a data rate of 155 Mb/s.

Thus, as we have seen, laboratories around the world are considering various techniques of providing packaging with low-cost alignment of optoelectronic components. As required component and system functionalities increase and pressures to reduce costs grow, these techniques will become increasingly important along with improved automated precision pick-and-place and robotic alignment tools [SLL94] for packaging of hybrid optoelectronic integrated assemblies.

CHAPTER ELEVEN

Free-Space Optical Interconnection for Digital Systems

JOHN NEFF

Optical fibers and planar integrated waveguides are now replacing electrical wires in digital systems that call for bandwidths in the gigahertz range and beyond. These guided interconnections, although well suited for communications of digital information over a limited number of channels, are problematic from a packaging and manufacturing viewpoint for systems requiring thousands of channels or more. Unguided (free-space) optics offers an exciting alternative to fibers and waveguides.

11.1 INTRODUCTION TO OPTICAL INTERCONNECTION

The system of interconnections by which the processing elements of parallel computers and the nodes of switching systems can share information among themselves is one of the most important characteristics of these digital systems, and unless the interconnections can handle data rates promised by photonics technology, future digital systems will be severely handicapped by their inability to pass enough information amongst the processing elements and/or switching nodes. Just as photonics is becoming the technology of the future for telecommunications, it will also impact communications within computers, especially for parallel systems. Not only do light waves have a much higher information capacity than electrical wires, but optical beams can pass through one another without interfering, leading to a higher packing density of interconnections.

11.1.1   The Importance of Optoelectronics for Interconnection

The demand for enhanced digital telecommunication and data communication services is creating a need for higher throughput—that is, higher channel bandwidth and a much higher density of channels. The emergence of parallel switching and computational architectures as solutions to this demand are being technologically constrained by conventional electrical interconnects. The parallel nature of these new architectures has shifted the burden for increased throughput from increasing the device speeds (conventional serial architectures) to increasing the throughput of the communications between the many parallel switching nodes (SNs) or processing elements (PEs). Some of the applications driving this demand are video on demand (VOD), large database machines (DBM), high-definition television (HDTV), control of large cellular communication networks, real-time graphics, weather and resource modeling, highly parallel communications with peripheral devices (e.g., archival memory), video conferencing, telepresence (e.g., remote surgery and health care monitoring), and 3-D display.

High switch/processor connectivity in a massively parallel switching fabric or computer is desirable for high switching or algorithmic efficiency yet difficult to implement electronically. The prevailing practice is to route messages through intermediate nodes, but this can lead to a very time-consuming process, especially for the larger parallel systems for which the average message path involves many intermediate nodes. A separate link between each switching node (SN) or processing element (PE) would lead to an unworkable mass of wires, each with its associated driver circuitry, all of which consume power and add weight. In addition, addressing and routing logic increases in complexity as the number of direct electrical connections increases. A tradeoff between switching/ algorithm efficiency and electronic complexity, power, and weight must be determined. This usually results in limited interconnect topologies such as mesh (nearest neighbor) or hypercube.

Free-space optical interconnects can ameliorate the inherent interconnect limitations of electronics. Light beams in free space do not require wires or fibers that add weight. They can intersect with no crosstalk and require less power for long, off-chip runs. Free-space optical interconnects have the additional advantage of easily traveling in any direction, utilizing all three physical dimensions, with a resultant increase in space utilization. However, probably foremost among the advantages of optical interconnects, especially those in free space, is the ability to implement interconnect topologies far more general than meshes and hypercubes.

Due to the uncharged nature of photons, optical interconnections potentially offer a freedom from mutual coupling effects. This greatly differentiates these interconnections from electrical ones based on charged carriers (electrons). Since the mutual coupling between electrical interconnects increases with increasing signal frequency, the advantage of optical interconnects becomes more important as the need for interconnect bandwidth increases. Although advanced dielectrics can overcome the electrical coupling effects at the lower frequencies, the frequency dependence of known dielectrics at frequencies approaching the gigahertz range leads to severe limitations for electrical interconnects at the bandwidths that will be required for implementation of high-throughput switching fabrics and parallel computers.

The uncharged nature of photons leads to another advantage for optical interconnects, that of lower power consumption for the longer interconnects. For electrical interconnects, electron charge leads to a distributed capacitance and resistance that creates an energy requirement which increases with increasing interconnect length. On the other hand, the energy requirement for optical interconnects lies with the optical source and the optical detector rather than being distributed along the interconnect. Therefore, beyond some distance (dependent on many link parameters), the optical interconnect becomes more energy-efficient [FEG88]. Since this break-even distance can be as small as subcentimeter, optical interconnects can lead to a considerable energy savings for large switching systems and massively parallel computers that will likely have hundreds of thousands of interconnects longer than a centimeter (or whatever the break-even distance happens to be).

There are numerous other advantages of optical interconnects that are noteworthy, such as the ability to frequency multiplex optical signals (due to the high bandwidth capacity of optical channels), the electrical isolation between optically interconnected electronic circuits, the increased fanout capability due to a freedom from capacitive loading effects, and a greater flexibility of routing due both to the ability of optical beams to pass through one another and to the freedom from routing signals in the presence of ground planes. However, it is the previously mentioned advantages of larger throughputs and lower energy requirements for the longer interconnects that figure most prominently into why optics is needed for the implementation of large parallel switching and computing systems.

Although fiber optics is being successfully applied to data communications between computers, labor-intensive assembly procedures for fiber systems will lead to free-space interconnections inside the computer switching fabric where thousands of channels are required. Upon consideration of the advantages of optical interconnects (both fiber-based and free-space) versus electrical interconnects, the free-space interconnects are best suited for channel lengths in the 1- to 10-cm range. From Figure 11.1, this is seen to include chip-to-chip, multichip module-to-multichip module, and board-to-board interconnections. The shorter interconnects will be electrical, while the longer ones will likely be fiber-based optical interconnects. It is free-space interconnection that will be the focus of this chapter.

11.1.2 The Evolution from Fiber-Based to Free-Space Optical Interconnection

Optical interconnection has been gradually evolving from the long-distance applications of city-to-city telecommunications to the very short-distance appli-

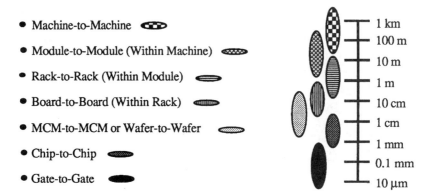

FIGURE 11.1. The hierarchy of computer interconnection.

cations within individual computers. The reason the evolution is moving in this direction rather than vice versa is twofold: (1) The advantage, and thus the market pull, of optics over electronics is greater for the greater distances and (2) the technology needed for short-distance interconnection is far more complex since large numbers of channels in a very limited area are usually required, and thus the integration of the electronics with the optics is of prime importance.

What can be called the first generation of optoelectronics for optical inter-connection involves serial fiber links as illustrated in Figure 11.2a. Such links take advantage of the large bandwidth of fibers to multiplex many electrical signals and transmit them optically by using the multiplexed signal to modulate a laser transmitter. The likely application for such links will be at the machine-to-machine level (e.g., local area networks) where the latency introduced by the multiplexing/demultiplexing operations will have little effect on overall system performance.

The second generation involves the use of laser or laser-modulator arrays, as illustrated in Figure 11.2b, thereby achieving high throughput rates without the need to multiplex. Although such links could be used in many of the same applications as the serial links, their advantage of lower latency makes them appealing for use further down in the interconnect hierarchy, such as module-to-module and rack-to-rack. Motorola has announced such a product (called the Optobus) that has been designed with low cost as a major consideration [Bur94].

The application of optical interconnection at levels of the interconnection hierarchy below rack-to-rack is awaiting development of technologies for achieving high levels of integration of the optical devices with the electronics. The realization of laser/modulator arrays integrated with arrays of logic circuits will launch the third generation of optical interconnection, as illustrated in Figure 11.2c. With the recent emergence of two-dimensional laser-modulator arrays (SEEDs from AT&T) and two-dimensional laser arrays made possible by vertical cavity surface-emitting laser (VCSEL) technology, very large numbers of optical channels (tens of thousands or more) are conceivable. This level of

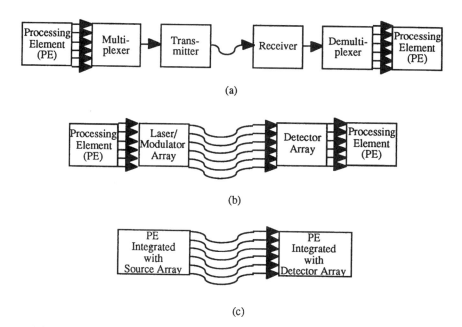

FIGURE 11.2. Three generations of optoelectronic interconnections [Wil93]: (a) First generation—serial fiber links; (b) second generation—parallel fiber links using discrete components; (c) third generation—integrated components, leading to free-space interconnections.

throughput has application at the board-to-board and MCM-to-MCM level of interconnection, where the design of digital electronic systems is feeling the constraint of the only hundreds of channels that are available from conventional electrical interconnects.

It is the third generation of optical interconnection based on two-dimensional arrays that leads to the need for free-space optics. As mentioned in the last section, labor-intensive assembly procedures that will be necessary to attach and align large numbers of fibers to optical sources and detectors will lead to free-space interconnections inside the computer or switching system where thousands of channels are required.

11.1.3 Applications for Free-Space Optical Interconnects

Applications requiring the communication of digital data on the order of a trillion bits per second (terabits/second or Tb/s) will require the high rates achievable with optical channels. There are two applications that are driving current developments in free-space optical interconnection: telecommunication/ data communication switching networks and fine-grained parallel computers. For the former, customer access to multimedia is projected to require the switching of hundreds of thousands of subscriber lines, each running at over

500 Mb/s [EYY93]. This results in throughputs that are three to four orders of magnitude beyond the capacity of existing telecommunication networks and one to two orders of magnitude beyond the projections for current electrical interconnect technology. In the case of fine-grained parallel computers, the need for tight coupling between tens of thousands of processing elements, each running near Gb/s data rates, also exceeds the projected capabilities of electrical interconnects.

Telecommunications/Data Communications

From a purely functional viewpoint, the crossbar switch system, which can facilitate the direct connection of any input channel to any output channel, is the ultimate in connectivity performance. Indeed, the crossbar switch can far out-perform buses under heavy traffic because of its ability to always establish a direct link between any inputs and any outputs. However, implementation of a crossbar system requires n^2 switches (where n is the number of input/output ports in the system). This is impractical for networks supporting even hundreds of customers, let alone thousands of customers or more. The number of switches can be significantly reduced by employing either space-division multiplexing, time-division multiplexing, frequency-division multiplexing, or some combination of these. Free-space interconnect solutions have been proposed for each of these multiplexing schemes.

A popular method for achieving space-division multiplexing (SDM) is to partition a large crossbar switch into many smaller crossbars, especially 2×2 crossbar switches. For example, n lines to be switched would be spread out over $(n/2)$ 2×2 crossbars, or they could be spread out over $(n/4)$ 4×4 crossbars, and so on. However, crossbars of less than n dimensions can only achieve switching between limited sets of lines rather than between all n lines. The full switching is achieved by using multiple stages of these small crossbars, as shown in Figure 11.3 for the 2×2 case. In order to realize the full switching of n lines, it is necessary to have $\log_2 n$ stages for 2×2 crossbars and $\log_4 n$ for 4×4

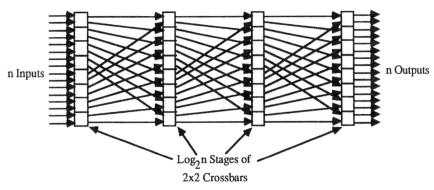

$\log_2 n$ Stages of 2×2 Crossbars

FIGURE 11.3. A multistage interconnection network (MIN). Reprinted with permission from [Nef96], copyright 1996 SPIE.

crossbars. Such switching networks are known as multistage interconnection networks (MINs). Their popularity comes from the reduced hardware complexity when compared to full crossbar switches while still maintaining the ability to establish a direct connection between any given input and any given output port. For a MIN employing 2×2 crossbars, $(n/2) \log_2 n$ switches are required, but this is significantly less than the n^2 switches needed for the full crossbar for large values of n. The cost and performance of MINs hit a reasonable balance between those of a bus and a crossbar. The application of free-space optics is for the stage-to-stage interconnections where n data paths are required between each stage and where some of these interconnects must traverse relatively long distances to achieve the proper connectivity between stages.

Time-division multiplexing (TDM) can be achieved via either of two alternatives: synchronous transfer mode (STM) or asynchronous transfer mode (ATM). STM is the conventional time-division multiplexing technique whereby each of m incoming signals has a transmission window of time length Δt every $m\Delta t$ seconds. This "bandwidth on assignment" can be inefficient if some of the signals are being transmitted less frequently than others. ATM overcomes this inefficiency by providing "bandwidth on demand." Instead of assigning the $m\Delta t$ time slots on a rotating basis, ATM assigns them on the basis of need, thereby ensuring that the channel is active as long as there is at least one signal to be transmitted. Broadband ISDNs (integrated services digital networks) for providing multimedia to the home are being designed around ATMs [FKK95]. Depending on how much processing is designed into an ATM switch in order to determine "need," large-scale ATM switches can involve several boards of electronics. Such multiboard systems may use free-space optics to overcome board-to-board interconnect bottlenecks.

Frequency-division multiplexing (FDM) has long been considered a good prospect for the application of optics, since the terahertz bandwidths of optical media are very appealing for frequency multiplexing. However, the large size of the hardware necessary to achieve a multitude of optical beams of differing wavelengths has restricted consideration of FDM to the upper levels of the interconnect hierarchy (Figure 11.1). But approaching on the horizon are two-dimensional, multiwavelength arrays of vertical cavity surface-emitting lasers (VCSELs) [CHZ91]. Such devices, if proven to be practical, will enable FDM free-space links on the board-to-board level.

A combination of SDM, TDM, and FDM is possible and will enable the throughputs that will be required in the not-so-distant future when millions of customers around the world will demand quick access to important databases (e.g., a universal medical diagnosis database) via desktop computers. Numerous governments around the world believe that a nation's power (both political and economic) in the twenty-first century will depend on that nation's information infrastructure and that data communication throughput rates at least in the tens of terabits (trillions of bits) per second will be needed. Such throughputs could be realized by a network with the relatively modest parameters of 100 time slots for TDM, 10 frequency slots for FDM, and 100 parallel channels, each operating at

200 Mb/s. This yields a throughput of 20 Tb/s, well out of the range of conventional electrical interconnection but quite suitable for optical interconnection.

Computers

Optical interconnects are gaining attention from computer architects who are designing parallel computers and are struggling with the constraints imposed on them by electrical interconnects. The power of parallel computers is directly related to the degree of communications between their processing elements, much as the success of a team of people working on a common problem is dependent on how well they share information with one another. The system of interconnections by which the processing elements can share information among themselves is one of the most important characteristics of any parallel computer. Unfortunately, the use of electrical interconnections is proving to be very limiting in the design of parallel systems. For example, when system architects scale up to more than a dozen or so processing elements, they have to abandon the preferred shared memory architecture in favor of the cumbersome technique of message passing due to the constraints imposed by interconnecting with electrical conductors. Parallel systems employing free-space optical interconnects will likely become more important with the trend toward processing the larger data fields associated with images, documents, sound, and video. Free-space optics, with its inherent parallel nature and with its capability of incorporating the phenomenon of beam steering, offers an architectural flexibility heretofore unseen in computer design.

At the personal computer end of the market, a massively parallel system can function as an accelerator for a host computer. Figure 11.4 illustrates planes of processing elements (called smart pixels when having optical I/O) mounted on a circuit board in such a way as to permit a large number of optical interconnect beams to travel from any one plane to its adjacent planes. Such an accelerator

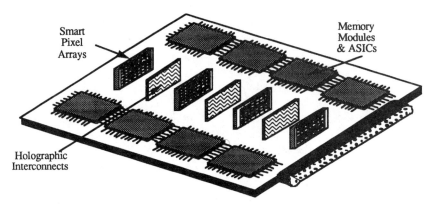

FIGURE 11.4. Concept for packaging an optoelectronic 3-D computer on an electronic printed circuit board. Reprinted with permission from [Nef 94], copyright 1996 IEEE.

would plug electrically into the host computer's backplane and would be used by the host to solve computational problems that could be mapped onto an array of identical processing elements (PEs). Such 3-D stacks of optically interconnected PE arrays (smart pixel arrays) are particularly suitable for implementing special purpose processors or accelerators for algorithms or processes that are amenable to pipelining. The power of such a system does not derive from fast data processing but rather from a very high degree of parallelism. Each PE array may be set up to perform a particular portion of the overall algorithm or process, passing its output optically to another PE array with a different processing capability. Figure 11.5 illustrates a couple of examples of such pipelining where each PE array has a different functionality. Figure 11.5a illustrates an image analysis type of problem in which the image pixels are passed from array to array, undergoing a different processing or transformation at each plane of the 3-D system. Communication of these pixels between the planes via a backplane would constrain performance. The same may be said for the problem domain illustrated in Figure 11.5b, that of image synthesis, in which an image is rendered

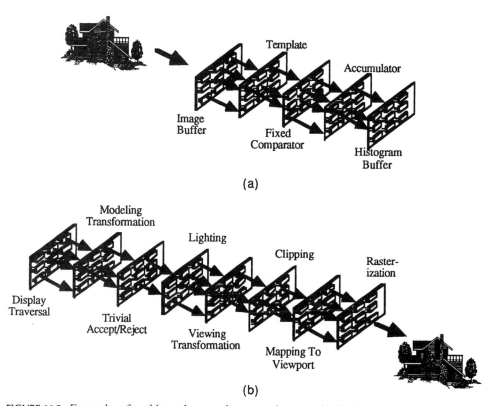

(a)

(b)

FIGURE 11.5. Examples of problems that may be mapped onto a pipelined architecture and therefore implemented as a 3-D system: (a) image analysis (e.g. histogram computation); (b) image rendering. Reprinted with permission from [Nef 96], copyright 1996 SPIE.

from a set of defining characteristics by undergoing a sequence of processing operations. This latter example represents the rapidly expanding area of computer graphics. Once again, large two-dimensional arrays of pixels are being processed in parallel and must be communicated between PE arrays having different functionalities.

The need for optical interconnects in parallel supercomputers has come to the forefront as a result of various programs around the world to develop computers capable of achieving sustained teraflops (FLoating point OPerations) performance. In the United States, this is the focus of the Federal HPCC program (High-Performance Computing and Communications), in which thousands of microprocessors are being interconnected to achieve performance at the teraflop level for general-purpose computing. Now that the teraflop program is established and proceeding according to plan, the U.S. government has set its sights on petaflop computing [SMS94], which is three orders of magnitude higher throughput than teraflop computing, or a million billion floating-point operations per second. This is very much a driver for massively parallel computing and optical interconnection. Table 11.1 lists just a few of the applications identified as needing such powerful parallel computers.

The value of free-space interconnection in the petaflop environment can be appreciated by considering one of the architectural scenarios that has been proposed for such systems, that of 10,000 100-gigaflop CPUs. For this scenario, something on the order of a 100,000,000,000 memory accesses per second will be required. Assuming a 64-bit word, this implies a 6400-Gb/s transfer rate between

TABLE 11.1. A Sampling of Application Areas Identified for Petaflop Computers

Design better drugs
Simulate functions of the human body
Understand the nature of new materials
Design fusion energy systems
Understand how galaxies are formed
Understand collision of black holes and gravitational waves
Perform high-level searches of full text databases
Forecast weather and predict global climate
Model flow of pollutants and ground water
Model ecological systems
Analyze planetary data to understand nature and use of land
Improve predictions of how and when earthquakes will occur
Build more efficient vehicles using computational fluid dynamics
Provide control and image analysis for advanced robots
Model economic conditions
Achieve dynamic scheduling of air and surface traffic when disrupted by weather and crisis
Produce graphics for digital movies
Support electronic shopping and other interactive television
Support worldwide digital library and information systems (text, images, video)

processor and memory, requiring 64 100-Gb/s optical interconnects. In addition, assuming that one out of 10 CPU instructions is input/output from/to a peripheral, implying a 640-Gb/s transfer rate, seven 100-Gb/s fibers would be required. Finally, assuming one out of 100 CPU instructions involves communicating with another processor, implying a 64-Gb/s transfer rate, an additional 100-Gb/s link would be required. If the computer is packaged as 10 MCMs/ board, 10 boards/card cage, and 10 card cages/rack, then assuming a factor of 10 increase in connectivity from rack to card cage to board to MCM, 64 100-Gb/s connections between processor and memory (rack to rack) implies 640 10-Gb/s links between card cages, 6400 1-Gb/s links between boards, and 64,000 100-Mb/s connections between MCMs. Since there are 10,000 CPUs (1 CPU/MCM), the above figures translate into $(64 + 7 + 1) \times 10,000 = 720,000$ 100-Gb/s optical links and $640 \times 10,000 = 6,400,000$ 10-Gb/s optical links, a real packaging and manufacturing nightmare. For an architectural scenario of 100,000 10-gigaflop CPUs (i.e. smaller PEs but more of them), the above reasoning leads to a requirement for 800,120 100-Gb/s links and 7,000,000 10-Gb/s links. For free-space implementation, one need not provide any physical guiding medium for each individual beam, thereby greatly simplifying the problem that will be faced in an attempt to interconnect such computing systems.

11.2 FREE-SPACE OPTICAL INTERCONNECTION ARCHITECTURES

Most of the envisioned systems for free-space optical interconnection fall into two classes: optical backplanes and 3-D systems. The former is generally seen as the more near-term of the two, employing free-space beams between optoelectronic arrays located on the ends of printed circuit boards in place of the multilevel electrical interconnect boards coming into use for advanced digital systems today. 3-D systems, on the other hand, distribute board-to-board optical interconnection over the extent of the boards, filling the dimension vertical to the boards with free-space beams.

11.2.1 Free-Space Optical Backplanes

The trend in digital systems, for the applications from personal computer accelerators to supercomputers that were discussed in Section 11.1.3, is toward teraflop computational platforms consisting of thousands of interconnected microprocessors. These systems will require backplanes that provide terabit-per-second throughput rates between a multitude of printed circuit boards (PCBs) containing high-performance chips and/or MCMs. These high throughput needs have stimulated the development of multilayered electrical backplanes consisting of stacked PCBs that collectively provide the board-to-board interconnections. But the throughput of these electrical backplanes is limited by parasitic inductance and capacitance and by resistive losses that severely limit the number of PCBs that can be stacked. Therefore, teraflop computational platforms will require new technologies that free them from the constraints of the conventional electrical interconnect technology.

A most promising technology for overcoming the limits of electrical backplanes is that of free-space optical interconnects. Tens of thousands of board-to-board data buses can be realized by optically interconnecting smart pixel arrays (PE arrays with optical I/O, as will be discussed in Section 11.3.1). These smart pixel arrays (SPAs) take the place of the metal connectors used in backplane systems today. One such architecture, illustrated in Figure 11.6, consists of point-to-point free-space channels between SPAs on adjacent PCBs [SzH95]. Each smart pixel serves as a node on one of the thousands of parallel buses that make up the optical backplane. In so doing, it must provide four basic capabilities: address recognition, injection, extraction, and regeneration. Although various levels of complexity can be incorporated into these capabilities and more capabilities can be added, a basic optical backplane can be realized by a pixel complexity on the order of only tens of electronic gates. This relatively low order of complexity is a major reason that the optical backplane is seen as being more near-term than 3-D systems, possibly realizable before the end of the century.

The physical backplane shown in Figure 11.6 is an optomechanical structure into which each board with its associated smart pixel arrays can be mounted in a rigid slot such that all pixels are inside the structure. Although the figure suggests that the regions between the smart pixel arrays are filled with air, they could

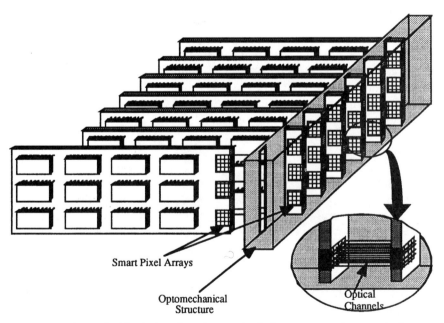

Smart Pixel Arrays

Optomechanical
Structure

Optical
Channels

FIGURE 11.6. An optical backplane based on optically interconnected smart pixel arrays. Reprinted with permission from [NeJ95].

consist of any transparent solid, thereby providing increased rigidity to the overall structure. In the interest of minimizing the complexity of the SPAs, the information flow via the free-space optical beams likely will be in only one direction, thus requiring a return path within the optomechanical structure for the array of light beams (not shown in the figure).

11.2.2 3-D Systems

Some of the applications discussed in Section 11.1.3, especially those involving image transformations such as those illustrated in Figure 11.5, may be easily mapped onto a massively-parallel, fine-grain (PEs of low complexity) computer that operates in the mode of single-instruction, multiple-data (SIMD) [Sto88]. Since such computers are massively parallel, there is a need for a very large number of interconnections between the many PEs. From a packaging view-point, such large numbers of interconnections are easier to accomplish via a technology such as free-space optics that does not require the physical connection of wires, fibers, or waveguides to drivers or sources. The fine-grained and single-instruction nature of such computers makes them particularly suitable to array implementation. With these characteristics, the circuits contained in the electronic chips in the optical backplane scenario of Figure 11.6 may be integrated into the smart pixel arrays as illustrated in Figure 11.7. The optical interconnects can now be seen as filling the area vertical to the boards—thus the name 3-D computer or 3-D system. The interconnect topology illustrated in Figure 11.7, with each node connected to all adjacent nodes in three dimensions

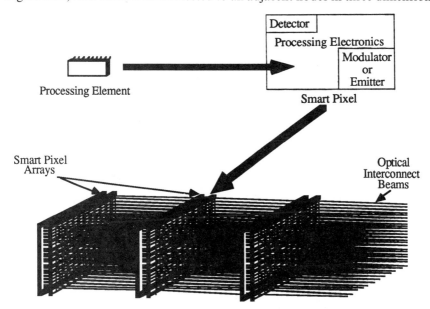

FIGURE 11.7. The evolution of the 3-D system from the SPA-based optical backplane. Reprinted with permission from [Nef 96], copyright 1996 SPIE.

(connected electrically in its plane and optically to adjacent planes), is known as a 3-D grid or mesh and is relatively low in complexity.

A major advantage of the 3-D system over the optical backplane is that the optical interconnect is generated in the vicinity of the signal source rather than at the edge of the board. As discussed below, this results in a reduction of signal delay. The major disadvantage is that only limited digital systems are prime candidates for being packaged in this 3-D format, whereas the optical backplane can be used for a wide range of multiple-board systems with high board-to-board throughput requirements. Another consideration in comparing the optical backplane to the 3-D architecture is the increased pixel complexity involved in 3-D systems, thereby leading to 3-D systems being viewed as somewhat longer-term than optical backplanes.

The use of direct board-to-board interconnects rather than backplanes results in lower signal delays, especially for multiprocessor systems that consist of a large number of nodes. A recent study [SMN95] has theoretically quantified this reduction in signal delay for the case of N nodes on one board being interconnected to N nodes on an adjacent board. Figure 11.8 shows the results via a plot of the average delay (averaged over all possible node pairs on adjacent boards) normalized by the signal propagation time between adjacent nodes (assuming all adjacent nodes are evenly separated). Also plotted for reference purposes are a curve for a 2-D mesh on a single board (not likely implementable for large numbers of PEs or nodes) and a curve for direct interconnection between all nodes (representing the physical limit for minimizing delay). As expected, the shorter the average length of the paths between communicating nodes, the smaller the signal delays. This is important since interconnect delay is the speed-limiting factor for most digital systems today.

Average delays can be reduced further than shown by the "3-D mesh using 3-D interconnects" curve of Figure 11.8 by taking advantage of the global

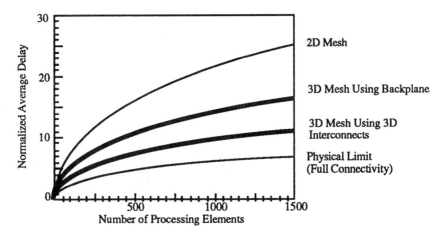

FIGURE 11.8. Communication delays for backplane versus 3-D interconnects.

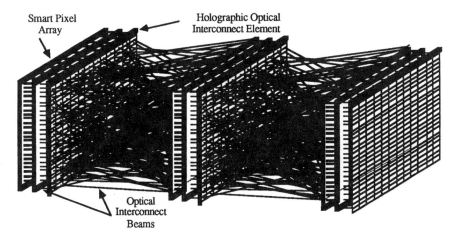

Smart Pixel
Array

Holographic Optical
Interconnect Element

Optical
Interconnect
Beams

FIGURE 11.9. A 3-D system based on holographic interconnects between stacked SPAs.

connectivity attribute of free-space optics. Mesh networks have been popular with electrical interconnect designers because they limit interconnects to the connection of nearest neighbor nodes, thereby avoiding a wiring nightmare with wires going all different directions. However, many node hops may be required to communicate a signal between a given pair of nodes, especially if the nodes are located in very different areas of their respective planes. Since free-space optical interconnects do not use any guiding media for the channels and since they do not suffer from the large degree of crosstalk between densely packed channels as do high-speed electrical interconnects, they are appealing for realizing a vast array of interconnect topologies between adjacent arrays. Also, since optics does not need additional energy for longer interconnect paths as does electronics with its distributed impedance effects, it is well within practical considerations to optically interconnect any node on one plane with any node on adjacent planes. This is known as global connectivity and is accomplished with free-space optical interconnects by using holographic optical elements (diffraction gratings) in the optical interconnect beam paths as illustrated in Figure 11.9. Each optical source on one plane has a corresponding holographic grating that can be fabricated to bend that source's optical beam in any direction. An array of such holographic gratings is known as a holographic optical interconnect element (HOIE). Such global connectivity can enhance performance for algorithms that involve global rather than local data relationships—for example, sorting (very prevalent in computation) and many transformations on data arrays such as fast Fourier transforms (FFTs) of images.

11.3 COMPONENTS FOR FREE-SPACE OPTICAL INTERCONNECTS

Although many optical interconnect links require conventional optical components such as lenses and mirrors, this section will focus on those components which have emerged relatively recently and, in so doing, have given birth to the field of free-space optical interconnection. These components are divided into two classes: smart pixel arrays (SPAs) and holographic optical interconnect elements (HOIEs).

11.3.1   Smart Pixel Arrays

Smart pixel arrays may be considered an extension of a class of optoelectronic components that have existed for over a decade, that of optoelectronic integrated circuits (OEICs). The vast majority of development in OEICs has involved the integration of electronic receivers with optical detectors and electronic drivers with optical sources or modulators. In addition, very little of this development has involved more than a single optical channel. However, OEICs have underpinned much of the advancements in serial fiber links (the first generation of optical interconnects illustrated in Figure 11.2a). SPAs encompass an extension of these optoelectronic components into arrays in which each element of the array has a signal processing capability. Thus, an SPA may be described as an array of optoelectronic circuits for which each circuit possesses the property of signal processing and, at a minimum, optical input or optical output. (Most SPAs will have both optical input and output.) Two other properties often associated with SPAs, especially two-dimensional arrays, but not required are free-space optical input/output (I/O) and identical processing in each pixel. The integration of the processing capability with the optical I/O is the enabling technology for the third generation of optical interconnects discussed in Section 11.1.2 and illustrated in Figure 11.2c.

An example of a smart pixel is shown in Figure 11.10. Note that this pixel has both electrical I/O and optical I/O. Such an interconnection capability suggests a

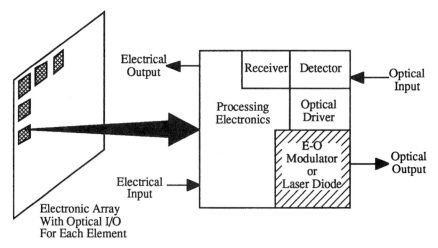

FIGURE 11.10.   Example of an element (pixel) of a smart pixel array. Reprinted with permission from [Nef 91], copyright 1996 SPIE.

utility as a critical component in systems that interface between the optical and the electronic realms (e.g., displays and digital video equipment). The electrical I/O of each pixel will most likely be for interconnection to neighboring pixels, but it could be for connection to buses integrated into the SPA to carry signals to and from the edge of the SPA. The latter would be the case if the SPA was a component, for example, of a display system employing row or column addressing.

Major Types of Smart Pixel Arrays

One defining characteristic of SPAs is the technology used for optical output. Some of these technologies will be mentioned here but not described. The interested reader may reference proceedings of topical meetings on smart pixel arrays [LEO92, LEO94]. Early development focused on optical modulators rather than laser diodes, since a good technology did not exist for realizing two-dimensional laser arrays, and much of this early effort dealt with modulator technology that is considered too slow (1–100 kHz) for interconnect applications. This includes modulators based on electrooptic effects in ferroelectric liquid crystals (FLCs) [JaJ89] and a ceramic containing lead, lanthanum, zinc, and titanium (PLZT) [LET86]. These electrooptic materials are bonded in some fashion to silicon circuits to create SPAs. An exception to the emphasis on slower-speed modulators has been the extensive effort at AT&T Bell Labs to develop high-speed modulators based on a quantum-well effect in gallium arsenide (GaAs) semiconductors [LHM89]. The modulators are known as self-electrooptic-effect devices (SEEDs), and devices incorporating field-effect transistors (FETs) with the modulators to realize SPAs are known as FET-SEEDs [Len94]. The difficulty of fabricating high-performance optical and electronic devices on the same GaAs substrate has spawned recent interesting work on bonding the GaAs modulators to silicon (Si) circuits to create hybrid GaAs-Si SPAs [Goo95].

A very appealing alternative to modulators for providing optical output for SPAs is the use of active sources. The major appeal of active sources over modulators is that the optical energy is created where it is needed rather than having to be directed in from some external source. The need for an external source leads to more components (larger overall system size), more opportunity for light scattering (due to more interfaces), and the need for fanout of the source beam into beamlets and the alignment of these beamlets onto the modulators. Early work in the area of active source arrays focused on light-emitting diodes (LEDs), because two-dimensional arrays of LEDs were easy to fabricate. However, LEDs suffer from a low external power efficiency and a long recombination lifetime (limited to modulation bandwidth). The fabrication of laser diodes, on the other hand, was limited to one-dimensional arrays because light emission was from the edge of the material structure rather than from the surface. It was not until the late 1980s that practical two-dimensional arrays of laser diodes became a reality when it was learned how to mass produce semiconductor structures that could lase vertically rather than horizontally

[JHS91]. These laser diodes are known as vertical-cavity surface-emitting lasers (VCSELs).

There are many efforts worldwide that are focusing on developing SPAs based on VCSEL arrays. VCSELs and light-emitting diodes (LEDs) overcome the shortcomings of optical modulators at the expense of increased heat dissipation. As mentioned above, VCSELs are preferred over LEDs because of higher external power efficiencies and higher operational frequencies (gigahertz rather than hundreds of megahertz). However, efficiencies are still too low for many applications, especially those requiring large arrays. For example, commercial-grade VCSELs currently being used in the Optoelectronic Computing Systems Center at the University of Colorado are capable of producing about 2 mW of optical output power when driven with 16 mA of current at 2.5 V ($\sim$5% power conversion efficiency). Research on VCSELs at many laboratories around the world is expected to result in a dramatic increase in power efficiency, thereby making VCSEL-based SPAs a practical alternative to those based on modulators. An efficiency as high as 50% has recently been reported by researchers at Sandia National Laboratories [LCS95], but such experimental devices are not yet ready for the commercial market.

The integration of optical modulators and sources with electronic devices has proved to be challenging. Silicon is well established as the material of choice for most electronic devices, but it is not optically active due to the indirect nature of its bandgap. GaAs is emerging as the "silicon" of optoelectronics, but even though GaAs electronic chips are readily available, the monolithic integration of electronic devices with optical modulators and sources has some disadvantages, a major one being the requirement for different GaAs structures for the optical devices than for the electronic devices if optimum performance is desired for both. In addition, the cost and levels of integration of GaAs electronics do not compare favorably with those of silicon electronics. For these reasons, much of the research and development effort on SPAs deals with hybrid rather than monolithic arrays. Figure 11.11 summarizes the advantages of hybrid versus monolithic SPAs. Section 11.4 describes the construction and packaging of VCSEL-based hybrid SPAs.

Quantitative Categories of Smart Pixel Arrays

SPAs may be quantified by three parameters, as illustrated in Figure 11.12: number of pixels, processing complexity per pixel, and optical I/O bandwidth. Physical limits associated with material and device processing dictate tradeoffs between these parameters. This means that the application will often determine these quantitative parameters. For example, an SPA fabricated for an optical backplane will likely trade off pixel complexity for a large number of pixels, whereas processing complexity per pixel is important for 3-D systems. And, although high I/O bandwidth is necessary for optical interconnect applications, SPAs fabricated for display applications will likely trade this off to gain a large number of pixels.

The majority of smart pixel arrays may be placed in one of three categories

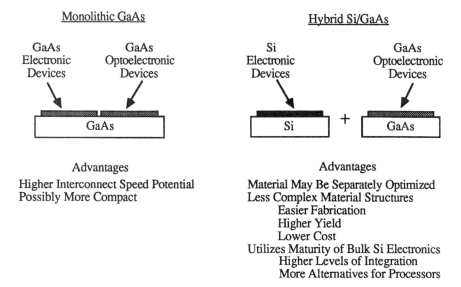

FIGURE 11.11.   Advantages of hybrid GaAs/Si versus monolithic GaAs smart pixel arrays.

according to their optical output: low-speed modulators (e.g., FLC and PLZT), high-speed modulators (e.g., SEEDs), and active sources (e.g., LEDs and VCSELs). Figure 11.13 suggests how these three categories might be mapped onto the 3-D parameter space, indicating which output mechanism is best suited

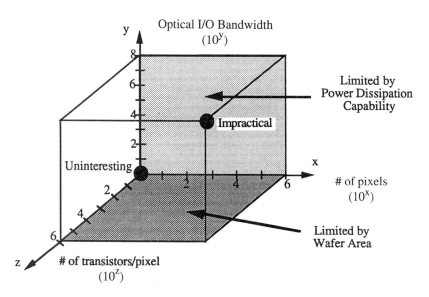

FIGURE 11.12.   3-D parameter space characterizing smart pixel arrays (courtesy of Ravi Athale, George Mason University).

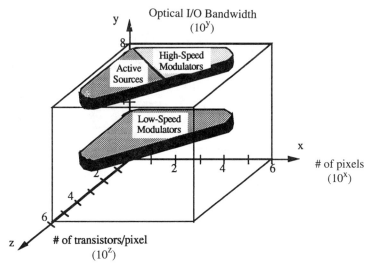

FIGURE 11.13. Suggested classification of the three major categories of SPAs according to their optical I/O bandwidth, number of pixels, and pixel complexity.

to the requirements of a particular application. SPAs based on low-speed modulators, although less expensive in general, are much better suited to display applications than computer interconnections because of their relatively low I/O bandwidth. With regard to higher-speed applications, the figure suggests that active sources and high-speed modulators might not be directly competitive as one might first conclude, but rather the selection of one over the other may be application-dependent. For example, an application requiring a large number of pixels with only a few transistors per pixel may call for a pixel-to-pixel pitch and thus an active source pitch—too small to effectively handle the required heat dissipation. Thus, modulators may be necessary for such applications. On the other hand, an application requiring a small number of complex pixels might be better served by active sources for the reasons cited above regarding the advantages of sources over modulators.

11.3.2   Holographic Optical Interconnect Elements

For optical interconnection, as for most optical systems, there is a need for beam-shaping and beam-directing elements such as lenses, mirrors, beam-splitters, and so on. The conventional means of achieving this has been to use materials that refract and/or reflect light. The problem with using refractive elements for microoptoelectronics is that relatively large three-dimensional volumes of transparent material are often required, resulting in bulky and heavy systems. Optical diffractive elements are finding increased use today [Fel94] in many optical systems because the refractive material can be replaced by planar elements that take up far less space and are much lighter in weight.

Thus, these elements are very important from a packaging viewpoint. They are a superposition of phase or amplitude gratings that bend each ray of an optical beam according to the periodicity and orientation of the grating through which that ray passes. In a collective sense, such ray bending can achieve any desired shaping or directing of an optical beam.

Diffractive optical elements can be realized by recording the pattern of two interfering optical beams—thus the name holographic optical elements (HOEs). For example, the diffractive element for a lens may be realized by making a hologram of that lens. But HOEs have come to be generated mostly by computer-generated patterns known as computer-generated holograms (CGHs) [Lee83]. The major advantage of computer generation of the pattern is that any transformation of a beam can be achieved as long as it can be expressed mathematically. For example, an aspheric lens HOE can be fabricated without using an actual aspheric lens, and wavefronts can be created even if the required physical refractive/reflective element does not exist. Also, the functions of several refractive optical elements can be combined into a single CGH (e.g., beam directing combined with beam collimation). In addition to weight and size advantages, CGHs have a lower unit cost than refractive optics since they can be mass produced with well-developed microelectronics technology; for example relief patterns can be etched from computer-generated masks [LoC89].

As illustrated in Figure 11.9, HOIEs are arrays of CGHs whose basic function is to bend each interconnect beam being emitted by one SPA toward a given detector or detectors on another SPA. If this is a point-to-point interconnection (no fanout), each subhologram of the HOIE is just a linear grating, as illustrated in Figure 11.14, such that the grating's periodicity determines the angle of deflection of the beam and its orientation determines the direction of deflection.

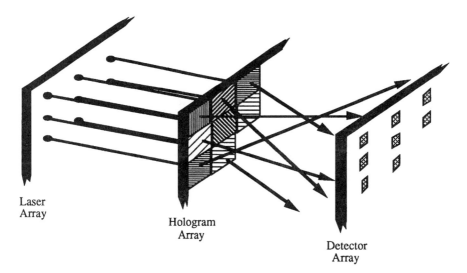

Laser Array

Hologram Array

Detector Array

FIGURE 11.14. Point-to-point beam-directing with an HOIE.

If any beam must be split into two or more beams (fanout > 1), that HOIE subhologram becomes a more complex grating with the dual function of beam splitting and beam directing.

Free-space interconnection requires more than just the point-to-point HOIE shown in Figure 11.14. First, each subhologram needs a well-collimated input beam rather than the diverging beam produced by an active source or reflected from a modulator. Second, the output beam from each subhologram must be focused onto the aperture of its intended detector. Both of these functions can be performed by refractive lenses, one preceding the subhologram and one following it. However, all three operations (collimating, directing, and focusing) can be combined into a single CGH as illustrated in Figure 11.15, thereby reducing the cost, size, and weight of the system, especially for large two-dimensional arrays.

Although HOIEs can consist of alternating regions of different optical absorption or different optical phase, the latter is more advantageous because optical energy is not wasted through absorption. Regions of alternating phase are most often achieved by varying the material thickness, thereby creating regions with different optical path lengths, leading to a varying phase across optical wavefronts passing through the material. Most HOIEs are fabricated by one of two processes: multistep photolithography or direct-writing with an electron beam. In the photolithography method, binary masks are produced from a data file created during the design process. The masks are then used sequentially to expose photoresist deposited on a transparent substrate, followed by an etch of the substrate in those regions unprotected by photoresist. The use of successively finer masks will produce a phase profile that comes closer and closer to the continuous-phase profile of refractive elements [LoC89], but this photolithographic process can be much cheaper and have much greater design flexibility than the grinding and polishing required for refractive element fabrication. Direct writing with an e-beam bypasses the mask process in that the e-beam is used to directly expose resist deposited on the substrate. In addition, varying the exposure of the e-beam can produce the many levels of the phase profile without the need for the often difficult alignment of binary masks [USL93].

The knowledge base underlying CGH fabrication is well established, but the lack of good computer-aided design (CAD) tools has inhibited the use of these

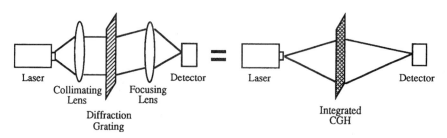

FIGURE 11.15. The combining of lens functions and beam-directing into a single CGH.

elements for most applications. The design of single-function CGHs is in itself very complex. The design of arrays of integrated CGHs interacting with optoelectronic devices would be almost an insurmountable task without the help of computers. Therefore, in order to make HOIE-based systems a reality, CAD tools need to be developed to perform as much of the design calculations as possible, and these tools should be integrated into a user-friendly CAD package. It will then be unintimidating for a systems engineer to create CGH designs, resulting in data files that can drive either photolithographic mask production or e-beam writing.

11.4 PACKAGING OF VCSEL-BASED FREE-SPACE INTERCONNECTS

A discussion of the packaging of free-space interconnect systems must address issues at the level of the SPA and at the systems level. Since SPA packaging is specific to the type of SPA, this discussion will focus on just one type of SPA, that involving the relatively new technology of VCSELs. Modulator-based SPAs involve some different issues, and the interested reader is referred to published literature on the AT&T work with their FET-SEED devices [Len94]. At the systems level, packaging focuses on how to arrange and align two or more SPAs so that they can exchange data over the optical links.

11.4.1 Smart Pixel Arrays

The VCSEL-based SPAs that will be discussed are hybrid components involving GaAs optoelectronic chips and silicon (Si) electronic chips. Although monolithic VCSEL-based SPAs are possible (electronic and optoelectronic devices on GaAs), limited work has been done in this area because of the difficulty in realizing a GaAs structure that can support the efficient operation of both electronic and optoelectronic devices. Creating a hybrid Si–GaAs structure involves epitaxially growing GaAs on Si or bonding the two together. Although the former is likely to lead to faster SPAs, it has proven to be a low-yield process because of the large lattice mismatch that exists between GaAs and Si, leading to unacceptable GaAs defect levels for fabricating laser diodes. A new bonding technique that won't be covered here but which looks promising, at least for GaAs detectors and LED sources, is called epitaxial lift-off (ELO) [Jok94a]. In this process, thin-film optoelectronic devices are grown on GaAs substrates but are then separated from the growth substrate using selective etching. These thin-film devices are then attached where desired to the silicon substrate and connected using standard microfabrication techniques. It has yet to be determined whether VCSELs can be bonded via ELO due to the sensitivity of VCSELs to material stresses that may be caused by the ELO process.

One way to combine the GaAs and Si chips is to mount both onto a common carrier which can support electrical microstrips between the two chips. The conventional way of doing this is to bond both to the carrier with their device sides up and then to electrically connect them by wirebonding both chips to the microstrips and their associated bonding pads on the carrier. For large array sizes, an unrealistic amount of space on the chips and on the carrier will be devoted to bonding pads, and the length of the electrical connections between the chips will defeat much, if not all, of the advantage of the optical interconnects. Borrowing a technique from the emerging technology of multichip module (MCM) fabrication, the chips can be placed device-side down (called flip-chip) and bump bonded to the carrier. Bump bonding has the distinct advantage that chip connections can be made anywhere on the surface of the chip rather than being confined to the chip's periphery, as is the case for wire bonding. This most often leads to a shortening of interconnect lengths, thereby enabling higher-speed operation. Furthermore, bump bonding can establish all chip connections in parallel, thus reducing production time for large arrays.

The flip-chip bonding of the optoelectronic chip to the common carrier leads to an important constraint on the carrier. Since the optical sources now face the carrier, it must be transparent to permit the optical beams to pass through to the outside of the hybrid structure. Glass is one material that can be used, but the superior thermal properties of sapphire make it very appealing from all but the cost viewpoint.

A packaged SPA based on the flip-chip bonding of both the VCSEL array and the electronic array (containing the processing elements and detectors) to a transparent carrier is shown in Figure 11.16. A hole is drilled in the well of a conventional ceramic package to allow passage of both the incoming and outgoing light beams. Note that the transparent substrate provides a convenient carrier on which to mount refractive and diffractive optical devices. Although

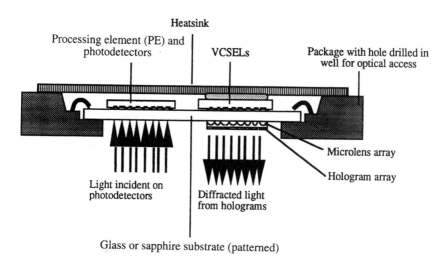

FIGURE 11.16. A packaging scheme for VCSEL-based hybrid smart pixel arrays. Reprinted with permission from [NML96], copyright 1996 SPIE.

only shown in the path of the outgoing beams, such beam-forming and directing devices could be used in the path of the incoming beams also (e.g., to focus the beams onto the detectors).

Future development will enable even tighter coupling between the GaAs and the Si chips. If the VCSEL chip is made to emit light in the opposite direction (i.e., in the direction of the substrate), it becomes possible to flip-chip this chip directly onto the Si chip, thereby minimizing the length of the electrical connections between the processing elements and the VCSELs (the two device surfaces are now facing one another). This is necessary if large-dimension, high-speed (gigahertz) SPAs are to be realized. Figure 11.17 illustrates one such pixel.

There are at least three possible ways to realize backside-emitting VCSEL arrays. First, the VCSELs can be made to emit in the wavelength region in which GaAs is transparent [CGC92]. This is the case for wavelengths longer than approximately 940 nm. (Most VCSELs are now being fabricated for wavelengths in the region of 830–840 nm.) Second, the common-wavelength VCSELs (830–840 nm) may be able to be grown on a substrate that is transparent at those wavelengths. Third, following the flip-chip bonding process, it may be possible to either remove or sufficiently thin the GaAs substrate to prevent significant absorption. These are all fertile areas for research and development.

Once substrate-emission VCSEL arrays are available, another problem remains to be solved before these arrays can be flip-chip bonded to silicon chips, that of heat removal from such a hybrid structure. Since light must exit from the backside of the VCSEL chip, a heatsink either must be transparent (e.g., diamond film) or must be attached to the silicon substrate and the VCSEL heat drawn to the silicon through the bonds. Until diamond films with close to 100% transparency are available, the latter approach is more acceptable. Initial estimates are that an 8×8 array of bump bonds will be capable of removing several watts of power dissipated by the VCSEL chips. At the present time, we are experiencing power dissipation levels for our 8×8 VCSEL arrays in excess of

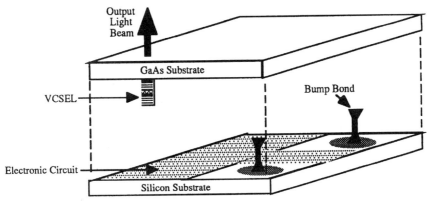

FIGURE 11.17. A smart pixel constructed by the flip-chip bonding of a VCSEL chip onto a silicon electronic chip. Reprinted with permission from [Nef 96], copyright 1996 SPIE.

2 W, but this is expected to decrease by at least an order of magnitude over the next few years. However, the 2-W figure is probably a realistic goal for SPAs, since increased efficiencies will likely be accompanied by increased array dimensions (e.g., 32 × 32 arrays). Another design problem must address whether heatsinks can adequately remove the combined heat of the VCSEL and silicon chips, including the VCSEL drivers. At the present time, our silicon-based VCSEL drivers are running at 10 MHz and dissipating about 15 mW each.

11.4.2   The Packaging of Multiple Smart Pixel Arrays

Most envisioned applications for free-space interconnects using SPAs require more than a single SPA. An exception would be optical interconnection between a single SPA and an optical memory disk, since the disk reflects light back to the SPA. The discussion of multiple-SPA systems will be divided into two categories: (1) two SPAs facing one another and (2) the stacking of many SPAs. Whereas the former is within reach today, the latter raises many interesting packaging questions that need answers before the systems can be engineered.

Two Facing Smart Pixel Arrays

Figure 11.18 illustrates two SPAs optically interconnected so as to pass two-dimensional data fields back and forth. The optical interconnects facilitate a global connectivity between the two SPAs that could be used to accelerate the operation of algorithms requiring relationships between data on a global scale (e.g., the popular FFT algorithm and various data-sorting algorithms). The host computer, upon needing the performance enhancement provided by such a globally interconnected processor, would upload the data to this optoelectronic unit for accelerated processing. Upon completion of its processing, the data would be downloaded back to the host computer. Since two facing SPAs and their associated optics can be placed in a miniature package, such an optoelectronic processor could be mounted on a board that would plug into the host's backplane. The Optoelectronic Computing Systems Center of the University of Colorado has designed and fabricated an optomechanical package for two 8 × 8 SPAs that is only 28 mm in diameter and 15 mm in depth [NeC95].

The ability to accurately model such optoelectronic systems is an important predecessor to the construction of these systems. Not only is the modeling important to avoid costly mistakes during the assemblage of 3-D systems, but it serves as the foundation for development of CAD tools that will enable systems engineers to design systems that will adequately meet their needs. The modeling determines important parameters for 3-D systems such as dimensions of the subholograms and detectors, Fourier lens parameters, impact of VCSEL wavelength variations, aberrations in the optical system, permissible packaging tolerances, magnitude of crosstalk and noise at the detectors, and bit error rate (BER). Several papers have been published or are in the process of being published that describe various aspects of 3-D system modeling [MTN95, MLN96, ZMN96].

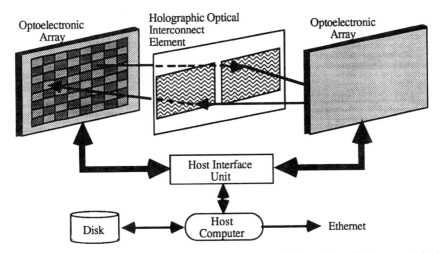

FIGURE 11.18. An optoelectronic interconnect system consisting of two facing smart pixel arrays. Reprinted with permission from [MTN95], copyright 1996 SPIE.

Stack of Multiple Smart Pixel Arrays

The use of only two SPAs would prevent their use in several of the application areas discussed earlier—for example, the pipelined architectures illustrated in Figure 11.5 and the optical backplane of Figure 11.6. These applications require several SPAs. The SEED-based systems of AT&T route the optical signals around the edges of the SEEDs, but this requires many additional optical components such as lenses, prisms, and beam splitters, adding cost, size, and weight [McC92]. The in-line stacking of SPAs as illustrated in Figure 11.19 will be challenging from at least two perspectives: (1) mounting heatsinks in such a way as not to block optical paths and (2) optically accessing detectors located on the front side of the SPAs (i.e., getting the optical beams through the SPA substrates and onto the detectors). A transparent heat conducting material such as diamond could solve the first problem. However, this technology is still in its infancy, and the design of systems on the assumption of its eventual availability and reasonable cost may pose an unacceptable risk. The second problem, that of addressing the detectors through the SPA substrate, requires making silicon appear transparent. One solution is to use longer wavelengths ($> 1\ \mu$m) for which silicon is transparent rather than absorbing, but VCSELs are not now available at these long wavelengths. Another interesting solution is to fabricate metal–semiconductor–metal (MSM) detectors on silicon membranes as illustrated in Figure 11.20 [LeV95]. The silicon substrate is etched from the bottom to within just a few microns of the MSM detector, thereby minimizing absorption in the silicon. Such detectors, addressable from both the front and backsides, have been demonstrated operating at speeds above 1 GHz at 830 nm.

The complications of stacking the SPAs can be overcome by arranging them in some compact way that avoids passing optical beams through the SPA

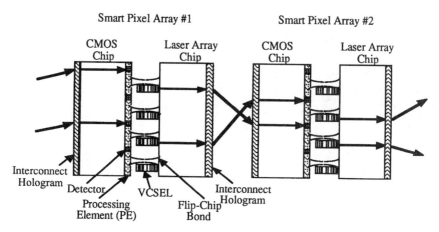

FIGURE 11.19. The in-line stacking of VCSEL-based smart pixel arrays.

substrates. One such solution is illustrated in Figure 11.21. A transparent slab of material with a parallel top and bottom is cut in such a way that the slab is a polygon when viewed from the top or bottom, as in Figure 11.21a. The SPAs can then be flip-chip bonded to the sides of this transparent polygon, thereby leaving their back sides available for attaching heatsinks. SPAs would alternate with HOIEs (operating in the reflection mode) on the polygon sides so as to facilitate optical interconnection between adjacent SPAs. That is, in order to achieve interconnection between two adjacent SPAs, an optical interconnect beam from one SPA would cross the transparent polygon to an HOIE on the other side and be reflected back across to the other SPA as shown in the figure. Not shown in the figure are the HOIEs integrated with each SPA that direct the light beams to the appropriate HOIE on the opposite side of the polygon.

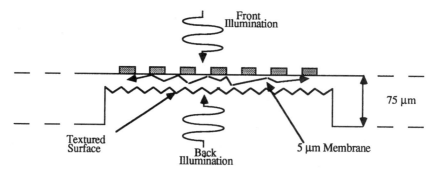

FIGURE 11.20. An MSM photodetector fabricated on an undoped silicon membrane created by etching the silicon substrate. The back surface of the membrane is roughened to enhance light trapping. The photodetector is capable of detecting light from either the front or back sides. Reprinted with permission from [Nef 94], copyright 1996 IEEE.

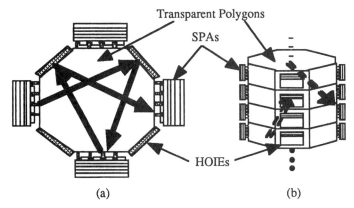

FIGURE 11.21. A packaging scheme for optically interconnecting multiple smart pixel arrays: (a) Smart pixel and holographic arrays mounted face down (flip-chip bonded) to the sides of a transparent polygon. (b) Stacks of transparent polygons optically interconnected via beams holographically directed from polygon to polygon.

Scaling of the system of Figure 11.21a is limited by the number of polygon sides. Scaling beyond this can be achieved by stacking polygons as shown in Figure 11.21b. The HOIEs can direct light beams out the top and bottom of the polygons as well as from side to side; thus, interconnect beams can travel between polygons in a stack. This would enable systems to be built with an arbitrary number of SPAs and would avoid the need for transparent heatsinks and substrates.

11.5 SUMMARY

There are two applications that are driving current developments in free-space optical interconnection: telecommunication/data communication switching networks and fine-grained parallel computers. For the former, customer access to multimedia is projected to require the switching of hundreds of thousands of subscriber lines, each running at over 500 Mb/s. This results in throughputs that are three to four orders of magnitude beyond the capacity of existing telecommunication networks and one to two orders of magnitude beyond the projections for current electrical interconnect technology. In the case of fine-grained parallel computers, the need for tight coupling between tens of thousands of processing elements, each running near gigabit-per-second data rates, also exceeds the projected capabilities of electrical interconnects.

Free-space optical interconnects offer an exciting alternative to conventional packaging for electronic systems which use planar board and backplane structures. The alternative involves using the space perpendicular to the processing planes for free-space optical beams that transport the data between the planes. Thousands of optical channels can be realized between the processing planes, either in the mode of an optical backplane or as what is known as a *three-dimensional system*. The major enabling technology for either of these modes is the two-dimensional optoelectronic (smart pixel) array.

A smart pixel array (SPA) for free-space optical interconnection would most likely consist of a two-dimensional array of identical processing circuits, each circuit being connected to an optical receiver and an optical transmitter (optical modulator or laser diode with its associated driver circuit). The functionality of the processing circuits will be very application-dependent. For example, an SPA for an optical backplane will likely perform the typical bus operations of signal extraction and injection, while one for image processing might be designed for image transformation (e.g., FFT or warping). What they have in common is the ability to input/output data at a very high throughput rate. Two SPAs can exchange data over thousands of channels, each running upwards of a gigabit-per-second data rate. This means that arrays of processing elements can exchange data at terabit-per-second rates, far in excess of electrically interconnected arrays.

This chapter discusses free-space optical interconnects from four perspectives: need, system architectures, enabling technologies, and component and system packaging. The need for such interconnection in telecommunication/data communication switching networks and fine-grained parallel computers is described. Since most of the envisioned systems fall into the two architectural classes of optical backplanes and 3-D systems, the chapter describes these classes in detail, and the reader is introduced to the enabling technologies of smart pixel arrays and holographic optical interconnect elements. Finally, the chapter describes attempts to implement free-space optical interconnect systems and some of the interesting packaging issues that these implementations raise.

CHAPTER TWELVE

Flip-Chip Assembly for Smart Pixel Arrays

Y. C. LEE, WEI LIN, and TIMOTHY McLAREN

Flip-chip assembly technologies are critical to packaging smart pixel arrays for free-space interconnects. These technologies can be categorized as solder assembly using tin–lead, gold–tin, indium, or other solder alloys or as solderless assembly using metal bumps or conductive adhesives. One of the main features of the solder technologies is self-alignment, which is important to a smart pixel array demanding accurate alignment between a device array and a substrate. This feature will be illustrated by a case study for a liquid crystal-on-silicon (LCOS) spatial light modulator. One of the main features of the solderless technologies is simplicity, which is important for an array demanding quick prototyping and/or low cost. This feature will be illustrated by a case study for a vertical cavity surface-emitting laser (VCSEL)-on-glass module.

12.1 INTRODUCTION

Smart pixel arrays, reviewed in the last chapter, are critical components for free-space interconnects. One of the main features of these arrays is the massive number of connections between an optical device and the associated electronic integrated circuits. If the arrays are assembled by hybrid packaging, it is desirable to use a one-step assembly process to make the thousands of connections. Flip-chip assembly is such a desirable technology.

Flip-chip assembly is moving into mainstream manufacturing. As shown in Table 12.1, the microelectronic chip pad count will increase significantly from 1995 to 2010 [SIA94]. In 1995, the number of I/O pads was 540 for the

TABLE 12.1. Chip Pad Count and Flip-Chip Technologies

Chip Pad Count	1995	1998	2001	2004	2007	2010
Commodity systems	208	256	324	420	550	700
Hand-held systems	300	450	675	880	1140	1500
Cost/performance systems	540	810	1200	1600	2000	2600
High-performance systems	900	1350	2000	2600	3600	4800
Automotive systems	132	200	300	400	500	700
Flip-chip (pitch of connections in mm)	0.250	0.200	0.150	0.100	0.100	0.100

cost–performance systems such as personal computers and will be over 2000 by the year 2007. Area-array connection technologies such as flip-chip are important to these high-I/O assemblies. By the year 2004, the pitch of the flip-chip connections should be reduced from 250 μm to 100 μm. These reduced pitches are for major microelectronics applications only. For many other applications such as chip-on-glass displays, the pitches are expected to be much smaller than 100 μm [Ada93].

IBM first introduced the flip-chip soldering technology three decades ago [Mil69]. Since then, the flip-chip assembly technologies have been driven by different applications. At the present time, there are six important application drivers: (1) multichip modules for mainframe computers, (2) single or a few chip packages for high-I/O CMOS VLSIs, (3) chip-on-glass for display driver chips, (4) prototypes of flip-chip assemblies, (5) optoelectronic modules, and (6) mixed signal modules. Driven by these different applications, different technologies have been developed for the flip-chip assembly.

The major categories of flip-chip assembly technologies are summarized in Figure 12.1 in terms of the key connecting materials. Flip-chip assembly using tin–lead solder is and will be the most popular approach. For optoelectronic packaging, gold–tin solder may be used because of its low-creep behavior, and indium solder may be used because of its compatibility with gold metallization. In addition, other lead-free alloys may become important when environmental policy is changed to add the special tax on the use of lead in electronics products. Most flip-chip solderless connection technologies have been developed to

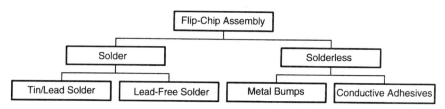

FIGURE 12.1. Flip-chip assembly technologies categorized by connecting materials.

support chip-on-glass for display driver chips. In that application area, conductive adhesives have been used widely. These applications require very fine pitch (e.g., 20 μm) but can tolerate high contact resistances. As a result, new, innovative flip-chip assembly technologies have been developed for them [LeW93].

Flip-chip assembly technologies can also be categorized by the connection mechanisms that are shown in Figure 12.2. These mechanisms are modified from the list of chip-on-glass technologies reviewed in [Ada93]. The first approach is flip-chip soldering using tin–lead, gold–tin, indium, or other alloys. Over the years, many advantages of this technology have been realized: high interconnection density, low cost, high yield, high strength, efficient heat conduction, batch assembly, and self-alignment during chip joining. In particular, its batch assembly and self-alignment features are critical to precision optical alignments. Through one solder reflow process, tens of thousands of flip-chip connections can be accomplished with an alignment accuracy less than 2 μm. These two features will be illustrated in more detail later in this chapter.

The remaining approaches are solderless alternatives. The second approach makes flip-chip connections through solid-state bonding. This approach is similar to wire bonding; however, it makes hundreds of connections by a single thermocompression or thermosonic bonding process. The bonding can be gold-bump-to-gold-pad bonding, gold bump-to-aluminum pad bonding, or other variations [KoS90, GoM92, KWM95]. The thermocompression bonding is accomplished by high temperature and pressure. In addition to these two parameters, thermosonic bonding applies ultrasound to soften the joint material and reduces the temperature and pressure required for good bonds. For example, for gold bump-to-gold pad bonding, the temperature can be reduced from 300°C to 150°C, and the force can be reduced from 100 gramf/joint to 30 gramf/joint when the material is softened by the ultrasound. Thermosonic

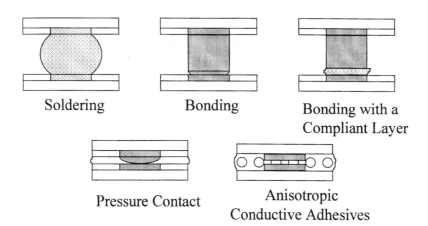

FIGURE 12.2. Flip-chip assembly technologies categorized by connecting mechanisms.

flip-chip bonding will be discussed in more detail later in this chapter. Thermosonic bonding is not used widely due to the complexity of the process; however, the parameters of this bonding process include those for thermocompression bonding.

Soft connecting material is desirable to accommodate the bump height variations. Ultrasonic energy can be used to soften materials, or a layer of compliant material can be used. The third approach uses such compliant materials (e.g., silver-loaded epoxy or indium) as the interface layer between the bumps and the pads [KKT93, MKS93, GTK95]. The compliant layer allows large deformations without demanding large pressing forces. For example, using the epoxy, the pressure required during bonding is reduced substantially to only 2 gramf/pad [Ada93]. The small force required can be applied by a simple pick-and-place machine; it also avoids device damage. However, there is a cost penalty associated with the additional layer, and the reliability of the soft connections under environmental changes is questionable.

The fourth and the fifth approaches accomplish the connections through electrical contacts rather than bondings. The fourth approach set the record on the fine-pitch flip-chip assembly [HFO90]. As shown in Figure 12.3, gold bumps with a 10-μm pitch and 3-μm height are formed on chips. The connections are accomplished by a compressive force resulting from the curing of the resin filled between the chip and the substrate. The resin is cured by UV light. With its 10-μm pitch connections and room temperature processing, this approach is the most advanced flip-chip connection technology to date. The technology was developed for mounting display driver chips in chip-on-glass applications. Because the original application for this process was not sensitive to the joint resistance, the reliability of the connections for smart pixel arrays with low resistance requirements should be studied carefully.

The fifth approach uses anisotropic conductive film (ACF) loaded with a single layer of metal balls to connect the chip and substrate I/O pads. Most of the balls used are made of a polymer coated with gold so that they are very compliant for high assembly yield and reliability. This approach is expected to be the main chip-on-glass technology due to its simplicity. However, its current carrying capability is limited by the thin metal film. There are alternatives with paste replacing the film, metal particles replacing the polymer balls, and other variations [LeW93]. However, ACF is still the preferred material for the chip-on-glass applications.

All of the flip-chip assembly technologies discussed above can accomplish a large number of connections with a fine pitch (below 200 μm). They are being developed to meet requirements for different flip-chip applications. It is impossible to single out an approach for the smart pixel arrays having a variety of assembly features. In fact, we expect to see several technologies used for these arrays. In order to understand the technologies in more detail, we will review two case studies. One is for soldering technology with an emphasis on self-alignment, and the other is for a solderless technology with an emphasis on thermosonic flip-chip bonding for low-cost prototyping and manufacturing.

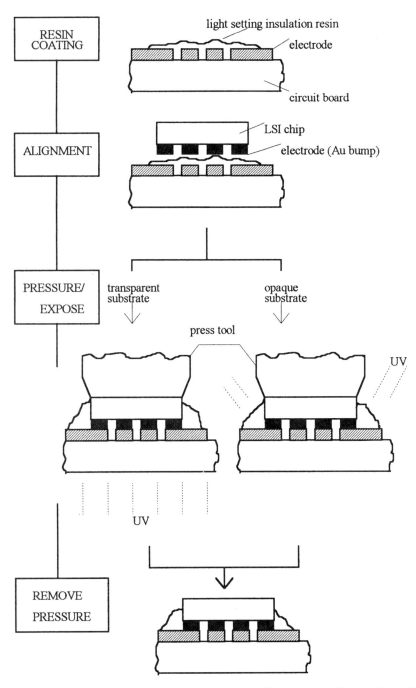

FIGURE 12.3. Process of microbump connections. Reprinted with permission from [LeW93].

12.2   SOLDER FOR LIQUID CRYSTAL-ON-SILICON

One of the main features of solder technology is the self-alignment mechanism illustrated in Figure 12.4. The molten solder wets the metal pads and forms a misaligned solder joint. The surface tension force moves the component to align with the substrate in order to minimize the energy of the solder joint. The final position is locked in by the solidified solder.

Flip-chip soldering technology is well known for its applications in microelectronics packaging. It is also being actively developed for optoelectronic packaging. There have been at least 68 publications reporting the use of flip-chip soldering for optoelectronic modules. They were reviewed by [TaL96] and [LeB94]. In general, these publications have documented a self-alignment accuracy of less than 2 μm with or without the use of mechanical stops/spacers. A review will not be repeated here. Instead, a case study will be presented to illustrate some major considerations.

The liquid crystal-on-silicon (LCOS) spatial light modulator (SLM) is critical to optical image processing and display. It can be packaged by a monolithic [KMJ96] or a hybrid approach [JaJ91]. The existing hybrid approach consists of several manual, sequential operations and should be improved. Solder technology was chosen for the improvement because of its batch and self-aligning assembly capability as well as its compatibility with manufacturing for flip-chip solder assembly [JLL95].

The schematic of a solder-assembled LCOS/SLM is shown in Figure 12.5 [JLL95]. The VLSI chip and the cover glass were directly soldered onto a silicon substrate. Spacers between the chip and the glass defined a gap to be filled with ferroelectric liquid crystal (FLC). The concave solder joint shape was designed to generate a force to pull the glass closely against the chip during solder reflow. In addition, the shrinkage of the solder joints in cooling applied another pulling force to control the gap.

To develop the solder-assembled LCOS/SLM, several critical considerations were (1) fabrication of solder pads, (2) solder material, (3) solder deposition, (4) solder reflow, (5) solder design, and (6) solder reliability. The details of these considerations were reviewed by [TaL96] and [LeB94]. In the following sections, we will discuss two considerations particularly important to the LCOS/SLM:

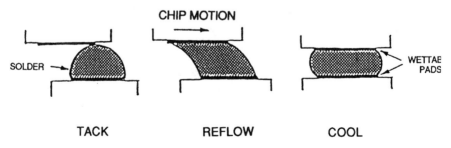

FIGURE 12.4. Flip-chip self-alignment soldering. Reprinted with permission from [LPL95].

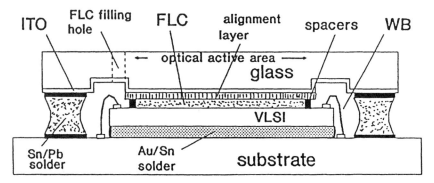

FIGURE 12.5. Detail structure of the soldering assembly LCOS SLM. Reprinted with permission from [JLL95], copyright 1995 IEEE.

solder design and reflow. In addition, the spacer fabrication and the assembly warpage will be briefly discussed.

12.2.1   Critical Considerations

Solder Joint Design
For better gap control and alignment accuracy, large surface tension forces were required. The force level was related to the solder joint design parameters such as pad size, pad geometry, joint height, solder volume, and the surface tension coefficient of the solder. For a good design at the micron level, a modeling tool was needed to estimate the restoring and pulling forces of the solder affected by different profiles [PSL92, Bra94]. The tool used to optimize solder parameters for the maximum pulling force for the LCOS/SLM was described in [LPL95]. The parameters include solder pad size, joint height, and solder volume applied. The solder pad size, fixed at 1.0-mm diameter, was constrained by the total modulator size limit. The 0.7-mm solder joint height, also fixed, was determined by the VLSI chip thickness. The optimization process was to decide how much solder needed to be applied for the maximum pulling force.

Using the modeling tool, the pulling force was calculated as a function of solder volume. As shown in Figure 12.6, as the solder volume is reduced, the joint shape changes from a cylindrical column to a concave profile, and the pulling force increases to a maximum. After that, if the volume is further reduced, the profile becomes too concave, causing the pulling force to drop. The optimum solder volume corresponding to the maximum force was 0.4 mm^3. The model-based design was important to assembling a high-quality LCOS/SLM [JLL95]. If the solder volume was chosen incorrectly (e.g., 0.6 mm^3), the force level could easily drop from 68 dyne to 22 dyne.

Fluxless Solder Reflow Process
The gap between the chip and the glass was in the micron level. If the gap were contaminated by liquid flux, it would be impossible to clean it. A fluxless,

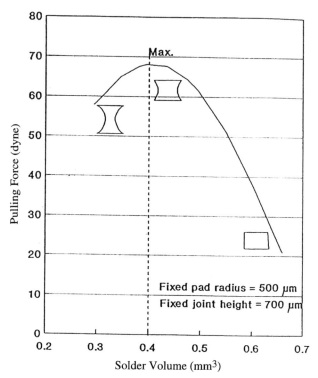

FIGURE 12.6. Optimum design of the solder volume for SLM assembly. Reprinted with permission from [LPL95].

residue-free soldering technology was developed for the LCOS/SLM assembly. There are two different approaches for fluxless soldering. The first one involves a precleaning process to remove surface oxide and a reflow process protected from oxidation. For example, PADS (plasma-assisted dry soldering) has a pretreatment to form a protective layer on the solder surface and results in efficient solder reflow in inert or even oxidizing ambients [KBN93].

The second approach uses reactive gas to remove solder oxides. By definition, the reactive gas is flux because its function is to remove surface oxides in order to enhance solderability. However, it is quite different from liquid flux. With proper control, the gas leaves little residue. Two kinds of reactive gases are commonly used in optoelectronic packaging: forming gas (4–10% H_2 in nitrogen or argon) [BTA93] and formic acid vapor (1.5–7% HCOOH in nitrogen) [Lin95, BrD93].

Figure 12.7 records self-alignment processes with respect to the use of forming gas (10% H_2 and 90% N_2) and formic acid vapor at different concentrations [Lin95]. The case with the forming gas was conducted at a reflow temperature of 280°C and assisted by the outgassing from a piece of FR-4 printed wiring board. The solder did reflow, but the self-alignment process was very slow. After 200 seconds, the component had not reached the steady-state, well-aligned

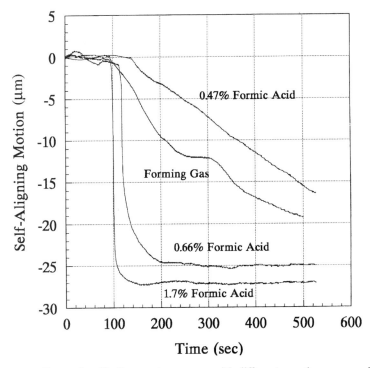

FIGURE 12.7. Dynamic self-alignment processes with different reactive gases and concentrations.

position. Similarly, low-concentration formic acid vapor was not effective. However, when the concentration reached 1.7% or higher, efficient self-alignment was completed within a few seconds. The reflow temperature with the formic acid vapor was only 220°C. Formic acid vapor resulted in efficient solder self-alignment, and this process was used to solder the LCOS/SLM.

Spacers and Warpage
In addition to the solder-related considerations, two other issues were critical to the LCOS/SLM assembly. They were the uniformity of spacer heights and the assembly warpage. Glass beads were often used as spacers. They can be applied by spraying with a random distribution onto the chip. However, some of them may land on the pixels. As the pixel arrays become larger (up to 1024 × 1024) and the size of each pixel becomes smaller, spacers may land on the pixels and degrade the image quality. As a result, a new spacer technology using photosensitive polyimide was developed [Lin95]. Polyimide spacers were accurately patterned in positions according to design requirements. They were fabricated by standard photolithography techniques. The spacer height was adjustable through the control of spin-coating speed. Submicron uniformity was routinely achieved.

Another important issue was the warpage of the chip and the substrate. When epoxy was used to attach the chip and the substrate to a package, the gap uniformity was poor due to bending caused by thermal expansion mismatches among the materials. Soft attachment material with low modulus such as silicone gel was used, reducing the bending dramatically [Lin95].

12.2.2 Summary

Solder self-alignment is important to smart pixel arrays demanding precision alignment. This feature was clearly shown in the case study for the solder-assembled LCOS/SLM. This technology can be used for other types of arrays, as detailed in [TaL96] and Chapter 8 of this book. Solder technology is popular in microelectronic and optoelectronic packaging; however, it is not simple. For many smart pixel arrays, solderless alternatives can be used for low-cost prototyping and manufacturing. Another case study will illustrate an example of such alternatives.

12.3 FLIP-CHIP BONDING FOR VCSEL-BASED SMART PIXEL ARRAYS

A smart pixel array consisting of VCSELs and CMOS detectors was developed for optically interconnected computing [MKZ95]. As shown in Figure 12.8, the 8 × 8 array of VCSELs on a GaAs chip was provided by Vixel Corporation with plated gold bumps 40 μm in diameter and 20 μm high. The detector array had

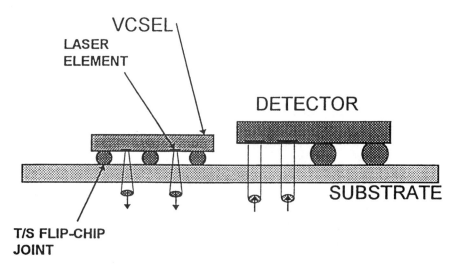

FIGURE 12.8. Cross-section schematic of a thermosonic flip-chip bonded optical transceiver. Reprinted with permission from [MKZ95], copyright 1995 IEEE.

aluminum contacts in an 8×10 bonding array on 250-μm pitch. Similar modules have been reported by Goossen et al. [GCJ93]; they soldered GaAs–AlGaAs multiple quantum well modulators onto an Si chip and then completely etched away the GaAs–AlGaAs substrate. Yeh and Smith [YeS94] attached an InGaAs quantum well VCSEL array onto both Si and Cu substrates. However, they used a gold contact surface on the VCSEL array and an In layer on the substrate. An Au–In alloy bond formed the joint, and the InGaAs substrate was then etched away. A different type of SLM has been made using epoxy for flip-chip bonding [JWL95]. This SLM consists of a hybrid of Si–PLZT with the Si substrate removed; final connections to the flip-chip are made after the chip bonding. Daryanani et al. [DFM95] bonded individual VCSEL elements using an InSn layer after first etching away the GaAs substrate from the array. Basavanhally and Brady [BaB94] developed a flip-chip assembly for SEED devices. The SEED device was reflow soldered to a quartz substrate using Pb–Sn eutectic solder.

To prototype the module shown in Figure 12.8, however, most of these technologies were not appropriate. It was difficult to deposit solder or any other materials onto the detector's aluminum pads because many steps would be required to process single dies for oxide removal, barrier metal deposition, and metal bumping. It was also not easy to deposit materials other than gold onto the VCSEL's gold pads. As a result, gold-bump-to-gold-pad bonding was chosen for the module prototyping [MKZ95].

The detector chips were gold-bumped using a ball bumping process on an automatic wire bonder with a special 1% Pd alloy gold wire made specifically for ball bumping. The VCSEL chip did not require any additional processing because the gold bumps were already plated. The glass substrate had a trace pattern with flip-chip bonding pads for the VCSEL and detector arrays and had a perimeter pad array for wire bonding to a package. The substrate metallization was 400 Å Ti/10000 Å Au on an 800-μm-thick fused silica glass substrate. The assembly shown in Figure 12.8 could be connected through one-step bonding, which was simple, dry, and compatible with the aluminum and gold metal pads for the VCSELs and CMOS detectors.

Flip-chip bonded assembly can be accomplished by the simultaneous deformation of all chip joints by using either thermocompression or thermosonic bonding. Table 12.2 lists the parameters that affect the bonding. Thermocompression bonding uses heat and pressure to soften and deform the joint material to form a bond. Both the chip and the substrate require heating to a temperature higher than 250°C, and an assembly pressure of around 1 N/joint is necessary to give adequate deformation for gold joints [KaL92]. Thermosonic bonding adds ultrasonic energy to induce acoustic softening of the joint material and reduces the temperature requirement on the substrate and significantly lowers the assembly force from that required for thermocompression bonding [KWM95]. High temperature and high pressure might damage the devices and the gold-to-aluminum interfaces; as a result, thermosonic flip-chip bonding was further developed for the module.

TABLE 12.2. Parameters Affecting Flip-Chip Bond Strength[a]

Parameter	Typical Values, Thermocompression	Typical Values, Thermosonic
Design Parameters		
Pitch	100–250 μm	100–250μm
Number of I/Os	3–300	3–300
Bump height	10–60μm	10–60 μm
Bump diameter	10–70 μm	10–70 μm
Chip contact material	Gold or Al	Gold or Al
Substrate contact material	Gold	Gold
Material thickness	1.0 μm	1.0 μm
Joint material	Gold	Gold
Process Parameters		
Substrate temperature	300–350°C	150–200°C
Chip temperature	200–350°C	Ambient
Bonding force	100 gf/bump	25–30 gf/bump
Bonding time	1–5 s	200–500 ms
U/S power	NA[b]	15–20 W
Bump height variation		
Coplanarity	< 0.1°	< 0.1°
Cleanliness		

[a] Modified with permission from [MKZ95], copyright IEEE, 1995.
[b] NA, not applicable.

12.3.1   Thermosonic Flip-Chip Bonding

The thermosonic bonding system is shown in Figure 12.9. The ultrasonic energy was provided by a power supply (Uthe Tech, Inc. 20G Ultrasonic Generator) used to energize the transducer (Uthe Tech, Inc. Model 29PT) [KWK93]. The ultrasonic bonding time was 500 msec, and the ultrasonic power output dial was set to 7.5 (approximately 75% of the rated 20 W full power). The assembly force on the VCSEL chip was 14.2 N (3.2 lb) and, on the detector chip, 18.7 N (4.2 lb). The substrate temperature was set to be 180°C.

Prior to assembly, it was important to ensure that the components were within specifications for the bonding pads and bumps. The bonding pads had to adhere strongly to the substrate, and the gold bumps had to be bonded/plated well to the aluminum–gold pads. Cleanliness of the bonding surfaces was also important to the consistency of the bonding [Jel77]. In addition, the end effector's planarity and placement accuracy had to be calibrated before the assembly.

During the assembly, it was necessary to control the temperature and the assembly force. A special end effector was designed to control the placement motion and the assembly force to avoid damage to the VCSEL [MKZ95]. More important, the impedance of the ultrasonic system had to be optimized. For the system shown in Figure 12.9, Kang et al. [KWL95] have shown that the motion

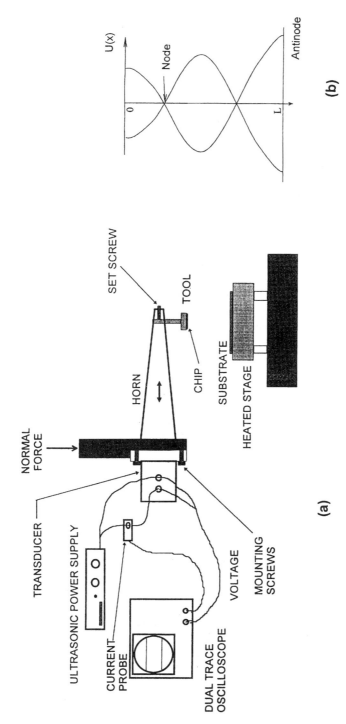

FIGURE 12.9. (a) Schematic diagram of the mechanical and electrical components of the ultrasonic flip-chip bonding system. (b) Typical diagram showing the motion amplitude at various locations down the length of the tool. Reprinted with permission from [MKZ95], copyright 1995 IEEE.

of the mechanical components of the ultrasonic system was directly related to the impedance of the electrical circuit. A listing of the parameters affecting the impedance is given in Table 12.3. The mechanical system was designed to operate near 60 kHz and was matched to the power supply. There was a phase-locked loop (PLL) circuit in the power supply that automatically adjusted the power supply frequency to the actual resonance frequency of the mechanical components. The power supply circuitry was designed to provide a fixed voltage output for a given power setting. Thus, the current was inversely proportional to the impedance. Minimizing the output circuit impedance maximized the output current and drive power. By monitoring the current (impedance), the effects of changes in the parameters could be observed. Prior to assembly, the parameters listed in Table 12.3 were adjusted so that the system impedance reached a minimum level in order to attain the maximum transfer of ultrasonic energy to the bonding site. Sufficient mechanical amplitude is required to induce acoustic softening in the gold joint material for the bonding deformation to take place.

With appropriate preparations for the preassembly and in-assembly processes, the thermosonic flip-chip bonding was in fact a very simple, one-step process. Sixty-four connections from a VCSEL and 80 connections from the CMOS detector to the glass substrate were each made simultaneously. The bonds were both electrically and mechanically good. There was no degradation of the VCSEL's performance. The bond strength was also better than the MIL-STD-883D standards: 5 gramf/joint [MKZ95].

Assembly Yield

Thermosonic flip-chip bonding is a simple process for quick prototyping. It is also important to low-cost flip-chip assembly. A major cost factor for the flip-chip solder assembly is the solder bumping process. By eliminating the need for solder bumping, this new bonding technology is able to reduce the assembly cost

TABLE 12.3. Factors Affecting Ultrasonic System Impedance (Motion Amplitude)

Factor	Comments
Electrical	
Frequency	Active PLL controlled for resonance
Phase	Set-up adjustment
Mechanical	
Tool length	Very sensitive, fixed by design
Tool mass	Sensitive, fixed by design
Tool position	Very sensitive, set-up adjustment
Tool orientation	With respect to rectangular shape of chip holder
Tool set screw torque	Checked periodically
Transducer mounting/screw torque	Set-up adjustment, set for maximum current

significantly. Unfortunately, its assembly yield is affected by not only ultrasonic energy transferred but also forces and coplanarity of the end effector, bump height variations, bump geometry, mechanical properties corresponding to different materials and temperatures, and distribution patterns of bumps.

Kang et al. [KXL93] developed an assembly yield model for thermocompression bonding. The model is also applicable to the thermosonic bonding which is really a thermcompression bonding process assisted by ultrasound. The model was run with different design and manufacturing parameter settings to obtain the minimum and maximum assembly forces that would provide 100% assembly yield for the joints based on the given criteria [MKZ95]. All of the trials used a bump height variation of 5%. Figure 12.10 shows the mean assembly force as a function of bump height, diameter, and assembly coplanarity angle for the 8 × 8 VCSEL array (64 I/O) with 250-μm pitch. From this chart we can conclude the following:

1. Assembly force decreases rapidly with bump diameter.
2. Planarity angle has little effect on mean force.
3. For the smaller bumps, the bump height has little effect on force.
4. Large-diameter, short bumps cannot tolerate planarity angles greater than 0.05°.

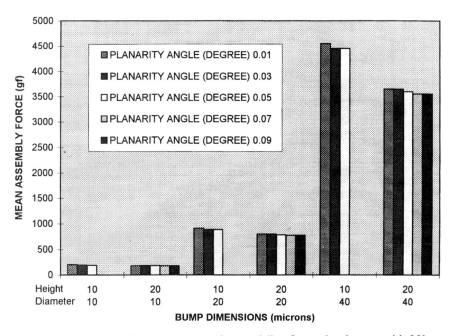

FIGURE 12.10. Results of thermocompression modeling for an 8 × 8 array with 250-μm pitching showing the mean assembly force required for a 100% yield. Reprinted with permission from [MKZ95], copyright 1995 IEEE.

The model developed is actually useful to almost all the solderless flip-chip assembly technologies presented in Figure 12.2. Using Figure 12.10, we can understand the similarities and differences among the thermosonic flip-chip bonding and other solderless technologies. If polymer balls or bonds with a layer of compliant material were used, the force required would be reduced by orders of magnitude. Such a reduction is critical to flip-chip assembly with hundreds or thousands of connections. The compliance also allows large deformations to accommodate large planarity angles. This is the reason why these two approaches are used widely for chip-on-glass applications. However, as mentioned earlier, they may not be acceptable for assembling every type of smart pixel array due to the limitation of the current-carrying capacity and reliability concerns.

For the very fine-pitch gold-to-gold contacts (the fourth approach shown in Figure 12.2) [HFO90], the small-diameter gold bumps also reduce the force requirement. However, they demand a higher level of control of the planarity because of the small bump height.

Clearly indicated by this brief analysis, there is not a single solderless flip-chip technology that is suitable for all smart pixel arrays. Different technologies are expected to be used to meet special requirements for a variety of arrays. Most technologies are moving into mainstream manufacturing for chip-on-glass applications, and they can be applied to smart pixel arrays without major modifications.

12.3.2 Summary

Thermosonic flip-chip bonding technology has been developed to prototype a smart pixel array consisting of VCSEL, CMOS/detector, and glass substrate. The technology is the flip-chip version of the widely used wire bonding technology. It is also as simple as wire bonding and has a potential to be used for the manufacturing of smart pixel arrays. A particular concern for the new technology is the effective transfer of the ultrasound from the transducer to the bonding site. Otherwise, most of the other considerations are similar to those for thermocompression bonding or other solderless assembly technologies. All these technologies are accomplished by effectively deforming the joints for bonding or contact. Based on different materials or connecting mechanisms, the solderless flip-chip assembly technologies will cover a wide spectrum of characteristics to meet the requirements of a variety of smart pixel arrays.

12.4 CONCLUSIONS

Flip-chip assembly technologies for smart pixel arrays have been reviewed with an emphasis on solder and thermosonic bonding technologies. Different solder and solderless technologies offer their own appealing features that may meet the special requirements for a particular type of smart pixel array. In general, self-alignment is the main feature associated with flip-chip solder assembly, and a simple process with a low force/temperature is the main feature associated with flip-chip solderless assembly. Most of the technologies reviewed are moving into mainstream manufacturing and will be available for prototype or manufacture of smart pixel arrays in the near future.

Acknowledgments
We would like to acknowledge support from the National Science Foundation (EDC-9015128 and MIP-9058409), ARPA-SAIC (N00014-93-C-0185), and Vixel Corporation (from U.S. Army contract no. DASG60-93-C-0052).

Summary of Acronyms and Abbreviations

ACF	anisotropic conductive film
ALO	aluminum-to-oxide bonding
APC	angled physical contact
APD	avalanche photodiode
AR	antireflection (also A/R)
ATM	asynchronous transfer mode
A/W	amps per watt
BGA	ball grid array
BISDN	broadband integrated services digital network
C4	controlled collapse chip connection bonding
CAD	computer-aided design
CATV	cable television
CD	core diameter
CGH	computer-generated hologram
CMOS	complementary metal oxide semiconductor
CSP	channel substrate planar
CR	clock recovery
CTE	coefficient of thermal expansion
CuPI	copper polyimide
CVD	chemical vapor deposition
CW	continuous wave
DACS	digital access and cross-connect system
dB	decibels
dBm	decibels relative to a power level of 1 mW
DCS	digital cross-connect system
DLD	dark line defect
DS-3	telephone multiplexed transmission rate standard of 1 Mb/s
DSO	Data Standards Organization

EBIC	electron beam-induced current
ECL	emitter coupled logic
EDP	ethylene diamine pyrocatechol
EEL	edge-emitting laser
ELO	epitaxial lift-off bonding
EMF	electromagnetic field
EO	electrooptic
ESCON	IBM standard connector for FDDI
ESD	electrostatic discharge
FAB	fiber array block
FBT	fused biconic technique
FDDI	fiber-distributed data interface (a standard)
FDM	frequency-division multiplexing
FET	field-effect transistor
FFT	fast Fourier transform
FHD	flame hydrolysis deposition
FLC	ferroelectric liquid crystal
FLOP	floating-point operation
FLOPS	floating-point operations per second
GaAs	gallium arsenide (an optically active semiconductor)
Gb/s	gigabits (one billion bits) per second
GHz	gigahertz (one billion cycles per second)
GI	graded index
HBT	heterojunction bipolar transistor
HDI	high-density integration
HDTV	high-definition television
HF	hydrofluoric acid
Ho	holmium
HOE	holographic optical element
HOIE	holographic optical interconnect element
HPCC	high-performance computing and communications
HR	high reflection
IC	integrated circuit
I/O	input/output
IR	infrared
ISDN	integrated services digital network
JVD	jet vapor deposition
kHz	kilohertz (one thousand cycles per second)
KOH	potassium hydroxide
LAN	local area network
LCOS	liquid crystal-on-silicon
LCP	left-circularly polarized
LD	laser diode
LED	light-emitting diode
LSI	large-scale integration

MAC	AT&T multifiber array connector
MAN	metropolitan area network
Mb/s	megabits per second
MCM	multichip module
MCM-D	dielectric MCM
MCM-L	laminated MCM
MFD	mode-field diameter
MFR	mode-field radius
MHz	megahertz (one million cycles per second)
MIN	multistage interconnection network
MM	multimode
MPa	megapascals
MSM	metal–semiconductor–metal
MTTF	mean time to failure
NA	numerical aperture
NAM	nonabsorbing mirror
NEP	noise equivalent power
nm	nanometers (one billionth of a meter)
NRZ	non-return to zero
ODL	optical data link
OE	optoelectronics
OEIC	optoelectronic integrated circuit
OFHC	oxygen-free high conductivity
PADS	plasma-assisted dry soldering
PBX	private branch exchange
PC	personal computer or physical contact (e.g., connectors)
PCB	printed circuit board
PD	photodetector
PE	processing element
PECVD	plasma-enhanced CVD
PI	polyimide
PIN	detector with an active intrinsic region sandwiched into a PN junction
PLL	phase-locked loop
PLZT	lead lanthanum zinc titanate, an electrooptic ceramic
PN	diode with a junction between a p- and an n-doped region
POF	plastic optical fiber
POLO	Parallel Optical Link Organization
RIE	reactive ion etching
RMS	root-mean squared
SAW	surface acoustic wave
SCI	scalable coherent interconnect (an IEEE standard)
SDH	synchronous digital hierarchy
SDM	space-division multiplexing
SEED	self-electrooptic-effect device (a type of optical modulator)

SEL	surface-emitting laser
SEM	scanning electron microscope
SIMD	single-instruction multiple-data
SLM	spatial light modulator
SM	single-mode
SMT	surface-mount technology
SN	switching node
SOA	semiconductor optical amplifier
SOH	section overhead
SONET	synchronous optical network
SPA	smart pixel array
SS	spot size
STM	synchronous transfer mode or synchronous transport module
SWQ-GRINSCH	single quantum well, graded index, separate carrier confinement heterostructure
Tb/s	terabits (one trillion bits) per second
TDM	time-division multiplexing
TE	transverse electric (a mode polarization)
TM	transverse magnetic
TO	thermooptic
USD	United States dollars
UV	ultraviolet
VCSEL	vertical cavity surface-emitting laser
VLSI	very large-scale integration (or VLSI chips)
WDM	wavelength-division multiplexing
WG	waveguide

References

[Ada93] K. Adachi, "Packaging Technology for Liquid Crystal Display," *Solid State Tech*, pp. 63–71 (January 1993).

[AlB89] D. A. Alles and K. J. Brady, "Packaging Technologies for III-V Photonic Devices and Integrated Circuits," *AT&T Tech J* **68**:1, pp. 83–92 (February 1989).

[And94] W. T. Anderson, "Consistency of Measurement Methods for the Mode Field Radius in a Single-Mode Fiber," *J Lightwave Tech* **LT-2**:2, pp. 191–197 (April 1994).

[Ani95] F. C. J. Anigbo et al., "Design of a Compact, High-Speed Optical Transceiver Using Two-Step Overmolding," *ECTC 1995 Proc*, p. 1104 (1995).

[ANT92] C. A. Armiento, A. J. Negri, M. J. Tabasky, R. A. Boudreau, M. A. Rothman, T. W. Fitzgerald, and P. O. Haugsjaa, "Four-Channel, Long-Wavelength Transmitter Arrays Incorporating Passive Laser/Single-Mode Fiber Alignment on Silicon Waferboard," *Proc ECTC'92*, p. 108 (1992); Conference on Optical Fiber Communications, p. 124 (1991).

[ARH96] A. Ambrosy, H. Richter, J. Hehmann, and D. Ferling, "Silicon Motherboards for Multichannel Optical Modules," *IEEE Trans Comp, Packaging, Manuf Tech Part A* **19**:1, p. 34 (1996).

[Arm91] C. A. Armiento et al., "Passive Coupling of InGaAsP/InP Laser Array and Single-Mode Fibers Using Silicon Waferboard," *Electron Lett* **27**, pp. 1109–1111 (1991).

[ASM89] *Electronic Materials Handbook, Vol. 1: Packaging* (ASM International Handbook Committee, 1989), Sections 6 and 9.

[ATJ91] C. Armiento, M. Tabasky, C. Jaganath, P. Haugsjaa, T. Fitzgerald, C. Shieh, V. Barry, M. Rothman, A. Negri, and H. Lockwood, "Passive Coupling of InGaAsP/InP Laser Array and Single-Mode Fibres using Silicon Waferboard," *Electron Lett* **27**, pp. 1109–1111 (June 6, 1991).

[ATK95] Y. Arai, H. Takahara, K. Koyabu, S. Fujita, Y. Akahori, and J. Nishikido, "Multigigabit Multichannel Optical Interconnection Modules for Asynchronous Transfer Mode Switching Systems," *IEEE Trans Comp, Packaging, Manuf Tech Part B* **18**:3, p. 558 (1995).

[BaB94] N. R. Basavanhally and M. F. Brady, "Opto-Electronic Packaging of 2-D Surface-Active Devices," *IEEE Lasers and Electro-Optics Society Conf Proc*, Boston, MA, pp. 215–216 (October 31–November 3, 1994).

[BAD91] N. S. Bergano, J. Aspell, C. R. Davidson, P. R. Trischitta, B. M. Nyman, and F. W. Kerfoot, "Feasibility Demonstration of Transoceanic EDFA Transmission Systems," *SPIE Fiber Laser Sources and Amplifiers III, Proc SPIE* **1581**, pp. 182–187, Bellingham, WA (1991).

[Bas78] E. Bassous, "Fabrication of Novel Three-Dimensional Microstructures by the Anisotropic Etching of (100) and (110) Silicon," *IEEE Trans Electron Dev* **ED-25**:10, pp. 1178–1185 (1978).

[Bas92a] N. Basavanhally, unpublished notes (1992).

[Bas92b] N. R. Basavanhally et al., "Optical Fiber Alignment Method," US Patent No. 5,135,590 (1992).

[Bas93a] N. R. Basavanhally, "Applications of Soldering Technologies for Optoelectronic Component Assembly," in *Advances in Electronic Packaging*, EEP Vol. 4-2 (New York: ASME, 1993).

[Bas93b] N. R. Basavanhally et al., "Optical Fiber Alignment Apparatus Including Guiding and Securing Plates," US Patent No. 5,185,846 (1993).

[Bas94] N. R. Basavanhally, "Alignment and Assembly Method," US Patent No. 5,281,301 (1994).

[Bas95a] N. Basavanhally, unpublished notes (1995).

[Bas95b] N. R. Basavanahlly et al., "Evolution of Fiber Arrays for Free Space Interconnect Applications," OSA Spring Topical Meetings (1995).

[Bas96] N. R. Basavanhally, "Apparatus for Aligning Optical Fibers in an $X-Y$ Matrix Configuration," US Patent No. 5,483,611 (1996).

[BaW92] N. R. Basavanhally and L. S. Watkins, "Opto-Mechanical Alignment and Assembly of 2D-Array Components," *Manufacturing Aspects in Electronic Packaging*, Winter Annual Meeting, EEP-Vol. 2/PED-Vol. 60, pp. 185–195 (New York: ASME, 1992).

[BBB96] N. R. Basavanhally, M. F. Brady, and D. B. Buchholz, "Optoelectronic Packaging of Two-Dimensional Surface-Active Devices," *IEEE Trans Components, Packaging, and Manufacturing Tech*, Part B: Advanced Packaging, **19**:1, pp. 107–115 (1996).

[BBF95] M. Berroth, W. Bronner, T. Fink, J. Hornugn, V. Hurm, T. Jakobus, K. Kohler, M. Lang, U. Howotny, and Z. Wang, "Monolithic Integrated Optoelectronic Circuits for Optical Links," *Technical Digest of the Optical Fiber Conference '95*, pp. 153–154 (1995).

[Bea78] K. E. Bean, "Anisotropic Etching of Silicon," *IEEE Trans Electron Dev* **ED-25**:10, pp. 1185–1193 (1978).

[BHN89] V. A. Bhagavatula, G. T. Holmes, D. A. Nolan, R. Jansen, and M. McCourt, "Planar Technology Enhances Coupler Performance," *Laser Focus World*, p. 155 (August 1989).

[BKL94] D. D. Bacon, A. Katz, C.-H. Lee, K. L. Tai, and Y.-M. Wong, "Permanent Metallic Bonding Method," U.S. Patent No. 5,234,153 (August 10, 1994).

[BKS80] R. A. Bergh, G. Kotler, and H. J. Shaw, "Single-Mode Fiber Optic Directional Coupler," *Electronics Lett* **16**:7, pp. 260–261 (1980).

[BMC94] B. L. Booth, J. E. Marchegiano, C. T. Chang, R. J. Furmanak, and D. M. Graham, "Polymer Waveguides for Optical Interconnect Applications," *Proc Conf on Optical Fiber Communications, 1994 Technical Digest Series* **4**, p. 74 (Washington, DC: Optical Society of America, 1994).

[Boo89] B. L. Booth, "Low-Loss Channel Waveguides in Polymers," *IEEE J Lightwave Tech* **7**:10, p. 1445 (October 1989) and Conference on Optical Fiber Communications, p. 63 (1989).

[Bou94] R. A. Boudreau, "Packaging for Optoelectronic Interconnections," *J Metals*, pp. 41–45 (June 1994).

[Bra94] K. A. Brakke, "Surface Evolver Manual," Version 1.94, Geometry Center, University of Minnesota, Minneapolis, MN (1994).

[BrC94] H. F. Breit and J. E. Cronin, "High Performance Ferrous/Non-Ferrous Metal Matrix Composite," *Texas Instruments Technical Library* (1994).

[BrD93] M. F. Brady and R. D. Deshmukh, "Solder Self-Alignment Methods," U.S. Patent #5,249,733 (1993).

[BrE90] H. Brugger and P. W. Epperlein, "Mapping of Local Temperatures on Mirrors of GaAs/AlGaAs Laser Diodes," *Appl Phys Lett* **56**:11, pp. 1049–1051 (March 12, 1990).

[BTA93] R. Boudreau, M. Tabasky, C. Armiento, A. Bellows, V. Cataldo, R. Morrison, M. Urban, R. Sargent, A. Negri, and P. Haugsjaa, "Fluxless Die Bonding For Optoelectronics," *IEEE Proc 43rd Electronic Components and Tech Conf*, pp. 485–490 (1993).

[Bub89] G. M. Bubel et al., "Mechanical Reliability of Metallized Optical Fiber for Hermetic Terminations," *J Lightwave Tech* **7**:10, pp. 1488–1493 (1989).

[Buc92] A. B. Buckman, *Guided-Wave Photonics* (Fort Worth: Saunders College Publishing, 1992).

[Bur94] D. Bursky, "Parallel Optical Links Move Data at 3 Gbits/s," *Electronic Design*, pp. 79–82 (Penton Publishing Inc., November 1994).

[CCB91] M. S. Cohen, M. F. Cina, E. Bassous, M. M. Oprysko, J. L. Speidell, F. J. Canora, Jr., and J. Defranza, "Packaging of High-Density Fiber/Laser Modules Using Passive Alignment Techniques," *IEEE Trans Photonics Tech Lett* **3**, p. 985 (1991).

[CCI90] CCITT Recommendation (International Telegraph and Telephone Consultative Committee), G.707 (1990).

[CFM95] S. H. Cho, S. Fox, S. A. Merritt, P. J. S. Heim, B. Gopalan, S. Kareenahalli, V. Vusirikala, C. E. C. Wood, and M. Dagenais, "Near-Ideal Diffraction-Limited Beam from a 970 nm High-Power Angled-Facet Tapered Semiconductor Optical Amplifier", in *Optical Amplifiers and Their Applications, 1995 OSA Technical Digest Series*, Vol. 18, pp. 24–27 (Optical Society of America, Washington, DC, 1995).

[CGC92] L. A. Coldren, R. S. Geels, S. W. Corzine, and J. W. Scott, "Efficient Vertical-Cavity Lasers," *Opt Quant Electr* **24**:2, pp. S105–S119 (1992).

[Che88] H. Z. Chen et al., "High-Frequency Modulation of AlGaAs/GaAs Lasers Grown on Si Substrate by Molecular Beam Epitaxy," *Appl Phys Lett* **52**:8, pp. 605–606 (February 1988).

[Che92] R. T. Chen et al., "Guided-Wave Planar Optical Interconnects Using Highly Multiplexed Polymer Waveguide Holograms," *J Lightwave Tech* **10**, pp. 888–897 (1992).

[Chi95] A. Chiou et al., "Photorefractive Spatial Mode Converter for Multimode-to-Single-Mode Fiber-Optic Coupling," *Optics Lett* **20**:10, pp. 1125–1127 (May 15, 1995).

[Cho95] H. K. Choi et al., "High-Power, High-Temperature Operation of GaInAsSb-AlGaAsSb Ridge-Waveguide Lasers Emitting at 1.9 μm," *Photonics Tech Lett* **7**:3 (March 1995).

[Chu93] Z. M. Chuang et al., "Photonic Integrated Tunable Receivers with Optical Preamplifiers for Direct Detection," *Appl Phys Lett* **63**:7 (August 16, 1993).

[CHZ91] C. P. Chang-Hasnain, J. P. Harbison, C. E. Zah, M. W. Maeda, L. T. Florez, N. G. Stoffel, and T. P. Lee, "Multiple Wavelength Tunable Surface-Emitting Laser Arrays," *IEEE J Quantum Electr* **27**, pp. 1368–1376 (1991).

[CMR92] P. C. Clements, R. Marz, A. Reichelt, and H. W. Schneider, "Flat-Field Spectrograph in SiO_2/Si," *Photonic Tech Lett* **4**, pp. 886–889 (1992).

[Coh91] M. S. Cohen et al., "Passive Laser-Fiber Alignment by Index Method," *IEEE Trans Photonic Tech Lett* **3**, pp. 985–987 (1991).

[Cou93] A. Coucoulas et al., "ALO Bonding: A Method of Joining Oxide Optical Components to Aluminum Coated Substrates," *Proc 43rd Electronic Components and Tech Conf*, pp. 470–481 (1993).

[Cow94] J. Cowles et al., "7.1 GHz Bandwidth Monolithically Integrated $In_{0.53}Ga_{0.47}As/In_{0.52}Al_{0.48}As$ PIN-HBT Transimpedance Photoreceiver," *IEEE Phot Tech Lett* **6**:8, p. 963 (1994).

[CPF95] J. Collins, R. Payne, C. Ford, A. Thurlow, I. Lealman, and P. Fiddyment, "Technology Developments for Low-Cost Laser Packaging," *Optical Fiber Conf Tech Dig*, pp. 222–223 (1995).

[Cro77] J. D. Crow et al., "GaAs Laser Array Source Package," *Opt Lett* **1**:1, pp. 40–42 (1977).

[CRS87] D. N. Christodoulides, L. A. Reith, and M. I. Saifi, "Theory of LED Coupling to Single-Mode Fibers," *J Lightwave Tech* **LT-5**:11 (November 1987).

[CSC94] R. Carson, P. Seigal, D. Craft, and M. Lovejoy, "Future Manufacturing Techniques for Stacked MCM Interconnections," *J Metals*, pp. 51–55 (June 1994)

[CWK91] W. Charczenko, P. S. Weitzsman, H. Klotz, M. Surette, J. M. Dunn, and A. R. Mickelson, "Characterization and Simulation of Proton Exchanged Integrated Optical Modulators with Various Dielectric Buffer Layers," *J Lightwave Tech* **9**:1, pp. 92–100 (1991).

[CYY95] A. Chiou, P. Yeh, C. Yang, and C. Gu, "Photorefractive Spatial Mode Converter for Multimode-to-Single-Mode Fiber-Optic Coupling," *Opt Lett* **20**:10, pp. 1125–1127 (1995).

[DaB94] M. Dautartas and A. Benzoni, "Design of a Low-Cost Self-Aligned MCM-D Based Optical Data Link," *Proc IEEE Laser and Electrooptics Society Annual Meeting*, pp. 71–72 (1994).

[DBW95] M. F. Dautartas, G. E. Blonder, Y.-H. Wong, and Y. C. Chen, "A Self-Aligned Optical Subassembly for Multi-Mode Devices," *IEEE Trans Comp, Packaging, Manuf Tech Part B* **18**:3, pp. 552–557 (1995).

[DCG94] F. P. Dabkowski, A. K. Chin, P. Gavrilovic, S. Alie, and D. M. Beyea,

"Temperature Profile Along the Cavity Axis of High Power Quantum Well Lasers During Operation," *Appl Phys Lett* **64**:1, pp. 13–15 (January 3, 1994).

[DDA94] G. Depestel, W. Delbare, K. Allaert, A. Ambrosy, T. Qingsheng, J. Vandewege, J. Verbeke, and M. Vrana, "Multifibre Electro-optical Modules Compatible with the Fibre in Board Technology," *Proc LEOS Annual Meeting*, pp. 224–225 (1994).

[DDF92] D. Danovitch, D. Dreps, E. Foster, R. Rizzo, C. Paddock, and W. Vetter, "Packaging the DM-1062 Gigabit Optical Module," *Proc LEOS 1992 Annual Meeting*, pp. 255–256 (1992).

[Des92] R. D. Deshmukh et al., "Active Atmosphere Solder Self-Alignment and Bonding of Optical Components," *Proc IEPS*, pp. 1037–1051 (1992).

[DFM95] S. Daryanani, H. Fathollahnejad, D. L. Mathine, R. Droopad, A. Kubes, and G. N. Maracas, "Integration of a Single Vertical-Cavity Surface-Emitting Laser onto a CMOS Inverter Chip," *Electron Lett* **31**:10, pp. 833–834 (May 1995).

[DiD87] R. W. Dixon and N. K. Dutta, "Lightwave Device Technology," *AT&T Tech J* **66**:1, p. 65 (1987).

[DMT82] E. Duda, A. Maumy, and C. Tonda, "System for Soldering a Semiconductor Laser to a Metal Base," U.S. Patent No. 4,321,617 (March 23, 1982).

[Dra91] C. Dragone, "An $N \times N$ Optical Multiplexer Using a Planar Arrangement of Two Star Couplers," *Photonic Tech Lett* **3**, pp. 812–815 (1991).

[DSM90] M. B. J. Diemeer, F. M. M. Suyten, and A. McDonach, "Photoinduced Channel Waveguide Formation in Nonlinear Optical Polymers," *Electron Lett* **26**:6, p. 379 (1990).

[DVV92] W. Delbare, L. Vandam, J. Vandewege, J. Verbeke, and M. Fitzgibbon, "Electrooptical Board Technology Based on Discrete Wiring," *Circuit World* **18**:3, pp. 11–16 (1992).

[DyD67] J. C. Dyment and L. A. D'Asaro, "Continuous Operation of GaAs Junction Lasers on Diamond Heatsinks at 200 K," *Appl Phys Lett* **11**:9, pp. 292–293 (November 1, 1967).

[Ebe89] K. J. Eberling, *Integrated Optoelectronics* (New York: Springer-Verlag, 1989).

[EPD93] C. A. Edwards, H. M. Presby, and C. Dragone, "Ideal Microlenses for Laser-to-Fiber Coupling," *J Lightwave Tech* **11**:2 (February 1993).

[EYY93] T. Egawa, K. Yukimatsu, and K. Yamasaki, "Recent Research Trends and Issues in Photonic Switching Technologies," *NTT Rev* **5**:1 (January 1993).

[FDB95] J. C. Feggeler, D. G. Duff, N. S. Bergano, C.-C. Chen, Y. C. Chen, C. R. Davidson, D. G. Ehrenberg, S. J. Evangelides, G. A. Ferguson, F. L. Heismann, G. M. Homsey, H. D. Kidorf, T. M. Kissell, A. E. Meixner, R. Menges, J. L. Miller, Jr., O. Mizuhara, T. V. Nguyen, B. M. Nyman, Y.-K. Park, W. W. Patterson, and G. F. Valvo, "10-Gbit/s WDM Transmission Measurements on an Installed Optical Amplifier Undersea Cable System," *Electr Lett* **31**:19, p. 1676 (1995).

[FEG88] M. Feldman, S. Esener, C. Guest, and S. H. Lee, "Comparison Between Optical and Electrical Interconnects Based on Power and Speed Considerations," *Appl Optics* **27**:9, pp. 1742–51 (May 1, 1988).

[Fel94] M. R. Feldman, "Diffractive Optics Move into the Commercial Arena," *Laser Focus World*, pp. 143–151 (October 1994).

√ [FFH79] K. Fujiwara, T. Fujiwara, K. Hori, and M. Takusagawa, "Aging Characteristics of $Ga_{1-x}Al_xAs$ Double-Heterostructure Lasers Bonded with Gold Eutectic Alloy Solder," *Appl Phys Lett* **34**:10, pp. 668–670 (May 15, 1979).

[FIF79] K. Fujiwara, H. Imai, T. Fujiwara, K. Hori, and M. Takusagawa, "Analysis of Deterioration in In Solder Used for GaAlAs DH Lasers," *Appl Phys Lett* **35**:11, pp. 861–863 (December 1, 1979).

[FKI92] M. Fujiwara, K. Kiyota, M. Itoh, and T. Uji, "An Optical Parallel Interface for Card-Cage to Card-Cage or Board-to-Board Optical Interconnections," Technical Report of The Institute of Electronics, Information and Communication Engineers, OE-92 (1992).

[FKK95] B. Furht, D. Kalra, F. Kitson, A. Rodriguez, and W. Wall, "Design Issues for Interactive Television Systems," *Computer* **28**:5 (May 1995).

[FLH95] W. Feng, S. Lin, R. B. Hooker, and A. R. Mickelson, "Study of Channel Waveguide Performance in Nonlinear Optical Polymer Films," *Appl Opt* **34**, pp. 6885–6893 (September 1995).

[FrM92] J. L. Freer and J. W. Morris, Jr., "Microstructure and Creep of Eutectic Indium/ Tin on Copper and Nickel Substrates," *J Electron Mater* **21**:6, pp. 647–652 (1992).

[FTA92] M. R. Feldman, I. Turlik, G. Adema, J. E. Morris, N. Hongzo, and W. H. Welch, *ECTC'92 Conference Proceedings*, pp. 513–518 (1992).

[Fuj82] K. Fujiwara, "Method of Mounting a Semiconductor Laser Device," U.S. Patent No. 4,360,965 (November 30, 1982).

[Fuk91] M. Fukuda, *Reliability and Degradation of Semiconductor Lasers and LEDs* (Artech House, 1991).

[Fuk95] M. Fukuda, personal communication (1995).

[FWB94] R. Fillion, R. Wojnarowski, W. Bicknell, W. Daum, and G. Forman, "Non-digital Extensions of an Embedded Chip MCM Technology," *Int J Microcircuits and Elect Packaging* **17**:4, pp. 392–400 (1994).

[GCJ93] K. W. Goossen, J. E. Cunningham, and W. Y. Jan, "GaAs 850 nm Modulators Solder-Bonded to Silicon," *IEEE Photonics Tech Lett* **5**:7, pp. 776–778 (July 1993).

[GDV91] G. Grand, H. Denis, and S. Valette, "New Method for Low-Cost and Efficient Optical Connections between Single-Mode Fibres and Silica Guides," *Electronic Lett* **27**, pp. 16–18 (1991).

[GhT89] A. Ghatak and K. Thyagarajan, *Optical Electronics* (New York: Cambridge University Press, 1989).

[Gia73] T. G. Giallorenzi et al., "Optical Waveguides Formed by Thermal Migration of Ions in Glass," *Appl Opt* **12**:6 (1973).

[GNH83] E. I. Gordon, F. R. Nash, and R. L. Hartman, "Purging: A Reliability Assurance Technique for New Technology Semiconductor Devices," *IEEE Electron Dev Lett* **EDL-4**:12 (December 1983).

[Gol69] L. S. Goldmann, "Geometric Optimization of Controlled Collapse Interconnections," *IBM J Res Dev* **13**, pp. 251–265 (1969).

[GoM92] C. E. Goodman and M. P. Metroka, "A Novel Multichip Module Assembly Approach Using Gold Ball Flip-Chip Bonding," *IEEE Trans Components, Hybrids and Manufacturing Tech* **15**:4, pp. 457–464 (August 1992).

[Goo95] K. W. Goossen et al., "GaAs MQW Modulators Integrated with Silicon CMOS," *IEEE Photonics Tech Lett* **7**, pp. 360–362 (1995).

[Gow93] J. Gowar, *Optical Communication Systems*, 2nd ed. (New York: Prentice-Hall, 1993).

[GPL92] G. J. Grimes, S. R. Peck, and B. H. Lee, "User Perspectives on Intrasystem Optical Interconnection in SONET/SDH Transmission Terminals," *Proc GLOBE-COM '92* (IEEE), Orlando, Vol. 1, pp. 201–207 (December 1992).

[GrH90] G. J. Grimes and L. J. Haas, "An Optical Backplane for High Performance Switches," *Proc Intl Switching Symposium* (IEEE), Stockholm, Vol. 1, pp. 85–89 (May 1990).

[Gri93] G. J. Grimes et al., "Packaging of Optoelectronics and Passive Optics in a High Capacity Transmission Terminal," *Proc 43rd Electronics Components and Tech Conf* (IEEE), Orlando, pp. 718–724 (June 1993).

[GTK95] J. L. F. Goldstein, D. B. Tukerman, P. C. Kim, and B. S. Fernandez, "A Novel Flip-Chip Process," *Proc Surface Mount Int Conf*, San Jose, CA, pp. 1–13 (August 1995).

[GWD95] K. Goossen, J. Walker, L. D'Asaro, S. Hui, B. Tseng, R. Leibenguth, D. Kossives, D. Dahringer, L. Chirovsky, and D. Miller, "4 × 4 Array of GaAs Hybrid-on-Si Optoelectronic Switching Nodes Operating at 250 Mbits/second," *LEOS Newsletter* **9**:1, pp. 21–25 (1995).

[Hah95] K. H. Hahn, "POLO—Parallel Optical Links for Gigabyte Data Communications," Proc. of the 45th Electronic Components and Tech. Conf., Las Vegas, Nevada, pp. 368–375 (1995).

[HaH73] R. L. Hartman and A. R. Hartman, "Strain-Induced Degradation of GaAs Injection Lasers," *Appl Phys Lett* **23**:3, pp. 147–149 (August 1, 1973).

[Hal94] S. A. Hall et al., "Assembly of Laser-Fiber Arrays," *J Lightwave Tech* **12**:10, pp. 1820–1826 (1994).

[Har89a] G. G. Harman, *Reliability and Yield Problems of Wire Bonding in Microelectronics* (ISHM, 1989), pp. 158–160.

[Har89b] D. H. Hartman et al., "Radiant Cured Polymer Optical Waveguides on Printed Circuit Boards for Photonic Interconnection Use," *Appl Optics* **18**:1, pp. 40–47 (January 1989).

[HaS94] B. L. Halpern and J. J. Schmitt, "Multiple Jets and Moving Substrates: Jet Vapor Deposition of Multicomponent Thin Films," *J Vac Sci Tech A* **12**:4, pt. 1, p. 1623 (1994).

[Hau95] P. Haugsjaa, "Optical Interconnects Between Multichip Modules," *Proc ISHM Workshop on Optoelectronics*, Ojai, CA (1995).

[HBK89] C. Henry, G. Blonder, and R. Kazarinov, "Glass Waveguides on Silicon for Hybrid Optical Packaging," *J Lightwave Tech* **7**:10, pp. 1530–1539, (1989).

[HBS93] W. R. Holland, J. J. Burack, and R. P. Stawicki, "Optical Fiber Circuits," *Proc 43rd Electronics Components and Technology Conf* (IEEE), Orlando, pp. 711–717 (June 1993).

[HDM96] P. O. Haugsjaa, G. A. Duchene, J. F. Mehr, A. J. Negri, and M. J. Tabasky, "Silicon Waferboard-Based Single-Mode Optical Fiber Interconnects," *IEEE Trans Components, Packaging, and Manuf Technol Part B* **19**:1, p. 90 (1996).

[HFJ78] K. O. Hill, Y. Fujii, D. C. Johnson, and B. S. Kawasaki, "Photosensitivity in Optical Fiber Waveguides, Application to Reflection Filter Fabrication," *Appl Phys Lett* **32**:10, p. 647 (1978).

[HFO90] K. Hatada, H. Fujimoto, T. Ochi, and Y. Ishida, "LED Array Modules by New

Technology Micro Bump Bonding Method," *IEEE Trans Components, Hybrids and Manuf Technol* **13**:3, pp. 521–527 (September 1990).

[HKO94] K. Hattori, T. Kitagawa, and M. Oguma, "Erbium-Doped Silica-Based Waveguide Amplifier Integrated with a 980/1530 nm WDM Coupler," *Electron Lett* **30**:11, p. 856 (1994).

[HOH93] Y. Hibino, H. Okazaki, and Y. Hida, "Propagation Loss Characteristics of Long Silica-Based Optical Waveguides on 5-inch Si Wafers," *Electron Lett* **29**:21, p. 1847 (October 14, 1993).

[Hor92] L. A. Hornak, ed., *Polymers for Lightwave and Integrated Optics: Technology and Applications* (New York: Marcel Dekker, 1992).

[HSY94] H. Hayashi, G. Sasaki, H. Yano, N. Nishiyama, and M. Murata, "Four-Channel Receiver Optoelectronic Integrated Circuit Arrays for Optical Interconnections," *IEICE Trans Electron* **E77-C**, pp. 23–29 (1994).

[HVM93] W. Hunziker, W. Vogt, and H. Melchoir, "Self-Aligned Optical Flip-Chip OEIC Packaging Technologies," *Proc ECTC*, pp. 84–91 (1993).

[HVM94] W. Hunziker, W. Vogt, H. Melchior, P. Buchmann, and P. Vettiger, "Passive Self-Aligned Low-Cost Packaging of Semiconductor Laser Arrays on Si Motherboard," *IEEE Photon Tech Lett* **7**:11, p. 1324 (1995); and W. Hunziker, W. Vogt, R. Hess, and H. Melchior, "Low-Loss, Self-Aligned Flip-Chip Technique for Interchip and Fiber Array to Waveguide OEIC Packaging," *IEEE LEOS '94 Conf Proc* **2**, p. 269 (1994).

[HYN95] H. Hsu, D. Yap, W. Ng, H. Yen, C. Armiento, M. Tabasky, J. Mehr, A. Negri, and P. Haugsjaa, "ARPA Analog Optoelectronic Module Program: Packaging Challenges for Analog Optoelectronic Arrays," *Proc 45th Electronic Components and Tech Conf*, pp. 565–596 (1995).

[IgO82] K. Iga, M. Oikawa et al., "Stacked Planar Optics: An Application of the Planar Microlens," *Appl Opt* **21**:19, pp. 3356–3460 (1982).

[IHI92] K. Iiyama, K. Hayashi, and Y. Ida, "Simple Method for Measuring the Linewidth Enhancement Factor of Semiconductor Lasers by Optical Injection Locking," *Optics Lett* **17**:16 (August 1992).

[Iml92] W. Imler et al., "Precision Flip-Chip Solder Bump Interconnects for Optical Packaging," *Proc 42nd ECTC*, pp. 508–512 (1992).

[IOC96] Integrated Optical Components, 3 Waterside Business Park, Eastways, Witham UK (1996).

[ITD95] M. Ivanov, T. Todorov, and V. Dragostinova, "Photoinduced Changes in the Refractive Index of Azo-Dye/Polymer Systems," *Appl Phys Lett* **66**:17, p. 2174 (1995).

[Ito91] M. Itoh et al., "Compact Multi-Channel LED/PD Array Modules Using New Assembly Techniques for Hundred Mb/s/ch Parallel Optical Transmission," *Proc 41st ECTC*, pp. 475–478 (1991).

[IYI91] S. Imamura, R. Yoshimura, and T. Izawa, "Polymer Channel Waveguides with Low Loss at 1.3 μm," *Electron Lett* **27**:15, p. 1342 (1991).

[Jac92] K. P. Jackson et al., "A Compact Multichannel Transceiver Module Using Planar-Processed Optical Waveguides with Flip-Chip Optoelectronic Components," *Proc 42nd ECTC*, pp. 93–97 (1992).

[Jac94] K. P. Jackson et al., "A High-Density Four-Channel OEIC Transceiver Module Utilizing Planar-Processed Optical Waveguides and Flip-Chip Solder Bump Technology," *AT&T Tech J* **66**:1, p. 1185 (1994).

[Jah92] J. Jahns et al., "Hybrid Integration of Surface-Emitting Microlaser Chip and Planar Optics Substrate for Interconnection Applications," *IEEE Photonics Tech Lett* **4**:12, pp. 1369–1372 (1992).

[JaH89] D. M. Jacobson and G. Humpston, "Gold Coatings for Fluxless Soldering," *Gold Bull* **22**:1, pp. 9–18 (1989).

[JaJ89] D. A. Jared and K. M. Johnson, "Ferroelectric Liquid Crystal Spatial Light Modulators," *SPIE Crit Rev Ser* **1150** (August 1989).

[JaJ91] D. A. Jared and K. M. Johnson, "Optically Addressed Thresholding VLSI/Liquid Crystal Spatial Light Modulators," *Opt Lett* **16**:12, pp. 967–969 (1991).

[JaM93] I. Januar and A. R. Mickelson, "Characteristics of S-Shaped Waveguide Structures by the Annealed Proton Exchange Process in LiNbO$_3$," *IEEE J Lightwave Tech* **11** (December 1993).

[JBD95] J. L. Jackel, J. E. Baran, A. d'Alessandro, and D. A. Smith, "A Passband-Flattened Acousto-Optic Filter," *Photonics Tech Lett* **7**:3, pp. 318–320 (1995).

[JCN94] C. Jones, K. Cooper, M. Nield, J. Ruch, J. Collins, I. Hall, A. McDonna, and S. Brown, "Hybrid Integration of Optical and Electronic Components on Silicon Mother-board," *Proc IEEE Laser and Electrooptics Society Annual Meeting*, pp. 273–274 (1994).

[Jel77] J. L. Jellison, "Kinetics of Thermocompression Bonding to Organic Contaminated Gold Surfaces," *IEEE Trans Parts, Hybrids and Packaging* **PHP-13**:2, pp. 132–137 (June 1977).

[JHS91] J. L. Jewell, J. P. Harbison, A. Scherer, Y. H. Lee, and L. T. Florez, "Vertical Cavity Surface-Emitting Lasers: Design, Growth, Fabrication, Characterization," *IEEE J Quantum Electr* **27**:6, pp. 1332–1346 (June 1991).

[JLL95] T. H. Ju, W. Lin, Y. C. Lee, D. J. McKnight, and K. M. Johnson, "Packaging of a 128 by 128 Liquid Crystal-on-Silicon Spatial Light Modulator Using Self-Pulling Soldering," *IEEE Photonics Tech Lett* **7**:9 (September 1995).

[JLT88] Special Issue on Integrated Optics, *IEEE/OSA J Lightwave Tech* **6**:6 (1988).

[JNL95] K. Jayaraj, T. Noll, and H. Lockwood, "A Low-Cost Hermetic Multifiber Array Packaging Technology," *Proc ISHM Workshop on Optoelctronics*, Ojai, CA (1995).

[JoD75] W. B. Joyce and R. W. Dixon, "Thermal Resistance of Heterostructure Lasers," *J Appl Phys* **46**:2, pp. 855–862 (February 1975).

[Jok94a] N. M. Jokerst, "Parallel Processing: Into the Next Dimension," *Opt Photonics News* (April 1994).

[Jok94b] N. Jokerst, "Thin-Film Multi-Material OEICs," *Conf Proc for LEOS 1994 Annual Meeting*, pp. 69–70 (1994).

[JPL87] JPL Invention Report NPO-16562/6062, "Deep, Precise Etching in Semiconductors," *NASA Tech Brief* **11**:3 (1987).

[JWL95] M. S. Jin, J. H. Wang, D. T. Lu, V. Ozguz, and S. H. Lee, "Direct Bonding and Flip-Chip Bonding Technologies Applied to Si/PLZT Spatial Light Modulator Fabrication," *IEEE Proc 43rd Electronic Components and Tech Conf*, Las Vegas, NV, pp. 194–200 (May 1995).

[Kae90] K. Kaede et al., "Twelve-Channel Parallel Optical Fiber Transmission Using a Low Drive Current 1.3 nm LWS Array and a PiN PD Array," *IEEE J Lightwave Tech* **8**:6, pp. 883–887 (June 1990).

[KaH77] B. S. Kawasaki and K. O. Hill, "Low-Loss Access Coupler for Multimode Optical Fiber Distribution Networks," *Appl Opt* **16**, pp. 1794–1795 (1977).

[KaL92] S. Y. Kang and Y. C. Lee, "Modeling of Flip-Chip Thermocompression Bonding: Part I—Physical Yield Models," *Manufacturing Aspects in Electronic Packaging, ASME EEP, Vol. 2/PED, Vol. 60*, pp. 147–163 (1992).

[KAM93] T. Kozaki, K. Aiki, M. Mori, M. Mizukami, and K. Asano, "A 156-Mb/s Interface CMOS LSI for ATM Switching Systems," *IEICE Trans Commun* **E76-B**:6, pp. 684–693 (1993).

[Kat92] T. Kato et al., "A New Assembly Architecture for Multichannel Single-Mode Fiber Pigtail LD/PD Modules," *Proc 42nd Electronic Components and Tech Conf*, pp. 853–860 (1992).

[Kaw90] M. Kawachi, "Silica Waveguides on Silicon and their Application to Integrated Optic Components," *Opt Quantum Electron* **22**, p. 391 (1990).

[KBN93] N. Koopman, S. Bobbio, S. Nangalia, J. Bousaba, and B. Piekarski, "Fluxless Soldering in Air and Nitrogen," *IEEE Proc 43rd Electronic Components and Tech Conf*, pp. 595–605 (1993).

[Ken92] D. L. Kendall et al., "Critical Technologies for the Micromachining of Silicon," in R. K. Willardson et al., eds., *Semiconductors and Semimetals*, Vol. 37, pp. 293–336 (San Diego: Academic Press, 1992).

[KeS74] D. B. Keck and P. C. Schultz, U.S. Patent No. 3,806,223 (1974).

[KGK96] K. Kurata, A. Gotou, N. Kitamura, S. Mizuta, S. Nakamura, and S. Ishikawa, "Hybrid Integrated Bidirectional Transmitter/Receiver Optical Module Based on Silica Waveguide Using Alignment-Free Hybrid Assembly Technique," *Tech Dig Opt Fiber Commun Conf* **2**, pp. 168–169 (1996).

[KHH94] H. Karstensen, C. Hanke, M. Honsberg, J. Kropp, J. Wieland, M. Blaser, P. Weger and L. Popp, "Parallel Optical Interconnect Modules with Multifiber Connectors," *Proc IEEE Electronic Components Tech Conf*, pp. 324–329 (1994).

[KKT93] T. Kusagaya, H. Kira, and K. Tsunoi, "Flip-Chip Mounting Using Stud Bumps and Adhesives for Encapsulation," ISHM MCM Conference, Denver, CO (April 14–16, 1993).

[KLT93] A. Katz, C.-H. Lee, K. L. Tai, and Y.-M. Wong, "Bonding Method Using Solder Composed of Multiple Alternating Gold and Tin Layers," U.S. Patent No. 5,197,654 (March 30, 1993).

[KMJ96] P. T. Kazlas, D. J. McKnight, and K. M. Johnson, "Integrated Assembly of Smart Pixel Arrays and Fabrication of Associated Micro-Optics," IEEE/LEOS Summer Topical Meeting on Smart Pixel Arrays, Keystone, CO (August 7–9, 1996).

[KMT95] S. Koike, S. Matsui, and H. Takahara, "Laser Diode Array Packaging for Coupling to Optical Polyimide Waveguides in Optoelectronic Multichip Modules," *Proc 45th ECTC*, pp. 766–769 (1995).

[Koi94] Y. Koike, "High-Speed Multimedia POF Network," *Proc POF '94* (IGI), Yokohama, pp. 16–20 (October 1994).

[KoM84] G. A. Koepf and B. J. Markey, "Fabrication and Characterization of a 2-D Fiber Array," *Appl Opt* **23**:20, pp. 3515–3516 (1984).

[KOO90] T. Kominato, Y. Ohmori, H. Okazaki, and M. Yasu, "Very Low-Loss GeO_2-Doped Silica Waveguides Fabricated by Flame Hydrolysis Deposition Method," *Electron Lett* **26**:5, pp. 327–329 (1990).

[KoS90] Y. Kondoh and M. Saito, "A New CCD Module Using the Chip on Glass Technique," *ISHM Proc* (1990).

[Koy93] K. Koyabu et al., "Novel High-Density Collimator Module," *OFC/IOOC '93 Tech Dig* (1993).

[KSM96] S. Koike, F. Shimokawa, T. Matsuura, and H. Takahara, "New Optical and Electrical Hybrid Packaging Techniques Using Optical Waveguides for Optoelectronic Multichip Modules," *IEEE Trans Comp, Packaging, Manuf Tech Part B* **19**:1, p. 124 (1996).

[KWB96] S. H. Kravitz, J. C. Word, T. M. Bauer, P. K. Seigal, M. G. Armendariz, "A Passive Micromachined Device for Alignment of Arrays of Single-Mode Fibers for Hermetic Photonic Packaging—The CLASP Concept," *IEEE Trans Comp, Packaging, Manuf Tech Part B* **19**:1, p. 83 (1996).

[KWK93] S. Y. Kang, P. M. Williams, T. A. Keyser, and Y. C. Lee, "Modeling and Experiment on Thermosonic Flip-Chip Bonding," ASME Winter Annual Meeting, New Orleans, LA (November 28–December 3, 1993).

[KWL95] S. Y. Kang, P. M. Williams, and Y. C. Lee, "Modeling and Experimental Studies on Thermosonic Flip-Chip Bonding," *IEEE Trans Comp, Packaging and Manuf Tech Part B*, pp. 728–733 (November 1995).

[KWM95] S. Y. Kang, P. M. Williams, T. S. McLaren, and Y. C. Lee, "Studies of Thermosonic Bonding for Flip-Chip Assembly," *Mat Chem Phys*, pp. 31–37 (1995).

[KXL93] S. Y. Kang, H. Xie, and Y. C. Lee, "Physical and Fuzzy Logic Modeling of a Flip-Chip Thermocompression Bonding Process," *ASME J Electronic Packaging*, pp. 63–70 (1993).

[KYT92] T. Kato, F. Yuriki, K. Tanaka, T. Haba, T. Shunura, A. Takai, K. Mizuishi, T. Tevaoka, and Y. Motegi, "A New Assembly Architecture for Multichannel Single-Mode Fiber Pigtail LD/PD Modules," *Proc ECTC '92*, pp. 853–860 (1992).

[KZC96] S. Kalluri, M. Ziari, A. Chen, V. Chuyanov, W. H. Steier, D. Chen, B. Jalali, H. Fetterman, and L. R. Dalton, "Monolithic Integration of Waveguide Polymer Electrooptic Modulators on VLSI Circuitry," *IEEE Photonics Tech Lett* **8**:5, p. 644 (1996).

[Lad93] I. Ladany, "Laser to Single-Mode Fiber Coupling in the Laboratory," *Appl Opt* **32**:18, pp. 3233–3236 (20 June 1993).

[LBN95] D. Leclerc, P. Brosson, R. Ngo, P. Doussiere, F. Mallecot, P. Gavigot, I. Wamsler, G. Laube, W. Hanziker, W. Vogt, and H. Melchior, "High Performance Semiconductor Optical Amplifier Array for Self-Aligned Packaging Using Si V-Groove Flip-Chip Technique," *IEEE Photonics Tech Lett* 7, pp. 476–478 (1995).

[LCG95] L. Lunardi, S. Chandrasekhar, A. Gnauck, C. Burrus, R. Hamm, J. Sulhoff, and J. Zyskind, "A 12-Gb/s High-Performance, High-Sensitivity Monolithic p-i-n/HBT Photoreceiver Module for Long-Wavelength Transmission Systems," *Photonics Tech Lett* 7, pp. 182–184 (1995).

[LCS95] K. L. Lear, K. D. Choquette, R. P. Schneider, Jr., S. P. Kilcoyne, and K. M. Geib, "Selectively Oxidized Vertical Cavity Surface-Emitting Lasers with 50% Power Conversion Efficiency," *Electr Lett* **31**, pp. 208–209 (February 2, 1995).

[LeB94] Y. C. Lee and N. Basavanhally, "Solder Engineering for Optoelectronic Packaging," *J Metals*, pp. 46–50 (June 1994).

[Lee83] S. H. Lee, Chairman/Editor, International Conference on Computer-Generated Holography, *Proc SPIE* **437** (August 1983).

[Len94] A. L. Lentine et al., "Field-Effect Transistor Self-Electrooptic-Effect Device (FET-SEED) Electrically Addressed Differential Modulator Array," *Appl Opt* **33**:14, pp. 2849–2855 (May 10, 1994).

[LEO92] *LEOS 1992 Summer Topical Meeting Digest on Smart Pixels*, IEEE Catalog #92TH0421-8 (IEEE Publishing Services, August 1992).

[LEO94] *LEOS 1994 Summer Topical Meeting Digest on Smart Pixels*, IEEE Catalog #94TH0606-4 (IEEE Publishing Services, July 1994).

[LET86] S. H. Lee, S. C. Esener, M. A. Title, and T. J. Drabik, "Two-Dimensional Silicon-PLZT Spatial Light Modulator," *Opt Eng* **25**, p. 250 (1986).

[LeV95] H. C. Lee and B. Van Zeghbroeck, "A Novel High-Speed Silicon MSM Photodetector Operating at 830 nm Wavelength," *IEEE Electr Dev Lett* **16**, pp. 175–177 (May 1995).

[LeW93] Y. C. Lee and P. M. Williams, "Solderless Connection Technologies," ASME Winter Annual Meeting, New Orleans, LA (November 28–December 3, 1993).

[LHM89] A. L. Lentine, H. S. Hinton, D. A. B. Miller, J. E. Henry, J. E. Cunningham, and L. M. F. Chirovsky, "Symmetric Self-Electrooptic-Effect Device: Optical Set-Reset Latch, Differential Logic Gate, and Differential Modulator/Detector," *IEEE J Quantum Electr* **25**:8, pp. 1928–1936 (August 1989).

[LiB95] Y. Liu and J. Bristow, "Optoelectronic Packaging and Interconnect Technology for MCM and Board-Level Applications," *Proc ISHM Workshop on Optoelctronics*, Ojai, CA (1995).

[Lin95] W. Lin, "Study of Soldering Assembly Technology for Liquid Crystal-on-Silicon (LCOS) Modules," Ph.D. Thesis, University of Colorado at Boulder (1995).

[LLP94] L. Lin, S. Lee, K. Pister, and M. Wu, "Micromachined Micro-Optical Bench for Optoelectronic Packaging," *Proc IEEE Laser and Electrooptics Society Annual Meeting*, pp. 219–220 (1994).

[LLW95] S. Lee, L. Lin, and M. Wu, "Micromachined Three-Dimensional Microgratings for Free-Space Integrated Micro-Optics," *Tech Dig 1995 Opt Fiber Commun Conf*, pp. 229–230 (1995).

[LoC89] J. Logue and M. L. Chisholm, "General Approaches to Mask Design for Binary Optics," *Proc SPIE* **1052**, pp. 19–24 (1989).

[LPL95] W. Lin, S. K. Patra, and Y. C. Lee, "Design of Solder Joints for Self-Aligned Optoelectronic Assemblies," *IEEE Trans Comp, Packaging and Manuf Tech Part A*, pp. 543–551 (August 1995).

[LRH94] I. F. Lealman, L. J. Rivers, H. J. Harlow, S. D. Perrin, and M. J. Robertson, "1.56 μm InGaAsP/InP Tapered Active Layer Multiquantum Well Laser with Improved Coupling to Cleaved Single-Mode Fibre," *Electronic Lett* **30**, pp. 857–858 (1994).

[LSR94] I. F. Lealman, C. P. Seltzer, L. J. Rivers, M. J. Harlow, and S. D. Perring, "Low Threshold Current 1.6 μm InGaAsP/InP Coupling to Cleaved Single-Mode Fibre," *Electron Lett* **30**, pp. 973–975 (1994).

[Mar91] D. Marcuse, *Theory of Dielectric Optical Waveguides*, 2nd ed. (Boston: Academic Press, 1991).

[Mas86] T. B. Massalski, "Binary Alloy Phase Diagrams," *Am Soc Metals* **2** (1986).

[MBW92] O. J. F. Martin, G.-L. Bona, and P. Wolf, "Thermal Behavior of Visible AlGaInP-GaInP Ridge Laser Diodes," *IEEE J Quant Electron* **28**:11, pp. 2582–2588 (November 1992).

[McC90] F. B. McCormick et al., "A Free-Space Cascaded Optical Logic Demonstration," Applications of Optical Engineering, *Proc SPIE* **1396** (1990).

[McC92] F. B. McCormick et al., "Experimental Investigation of a Free-Space Optical Switching Network Using Symmetric Self-Electrooptic-Effect Devices," *Appl Opt* **31**:26, pp. 5431–5446 (September 10, 1992).

[McC94] F. B. McCormick et al., "155 Mb/s Operation of a FET-SEED Free-Space Switching Network," *IEEE Photonics Tech Lett* **6**:12, pp. 1479–1481 (December 1994).

[Mer92] K. Mersereau et al., "Fabrication and Measurement of Fused Silica Microlens Arrays," in Miniature and Microoptics, *Proc SPIE* **1751**, pp. 229–233 (1992).

[Mer95] S. A. Merritt et al., "Measurement of the Facet Modal Reflectivity Spectrum in High Quality Semiconductor Traveling Wave Amplifiers," *IEEE J Lightwave Tech* **13**:3, pp. 430–433 (March 1995).

[MHC95] S. A. Merritt, P. J. S. Heim, S. Cho, and M. Dagenais, "A Reliable Die Attach Method for High Power Semiconductor Lasers and Optical Amplfiers," *Proc 45th Electronic Components and Tech Conf* (ECTC), Las Vegas, Nevada, pp. 428–430 (May 21–24, 1995).

[Mic93] A. R. Mickelson, *Guided Wave Optics* (New York: Van Nostrand Reinhold, 1993).

[Mid79] J. E. Midwinter, *Optical Fibers for Transmission*, Chapter 13 (New York: John Wiley & Sons, 1979).

[Mid93] J. E. Midwinter, ed., *Photonics in Switching*, Vol. 1 and 2 (New York: Academic Press, 1993).

[Mil69] L. F. Miller, "Controlled Collapse Reflow Chip Joining," *IBM J Res Dev* **13**, pp. 239–250 (1969).

[Mil78] C. M. Miller, "Fiber Optic Array Splicing with Etched Silicon Chips," *Bell Syst Tech J* **57**:1, pp. 75–90 (1978).

[Mil88] C. M. Miller, "Fiber Components and Interconnections," Conference on Optical Fiber Communications, Tutorial, p. 231 (1988).

[Mil89] D. A. B. Miller, "Optics for Low-Energy Communication Inside Digital Processors: Quantum Detectors, Sources and Modulators as Efficient Impedance Converters," *Opt Lett* **14**:2, p. 146 (1989).

[MKH94] G. R. Mohlmann, M. K. Kleinkoerkamp, J.-L. P. Heideman, T. A. Tumolillo, Jr., M. Van Rheede, C. Ramsamoedj, C. P. J. M. van der Vorst, and R. A. P. van Gassel, "Optical Polymers and Related Photonic Devices," *Proc SPIE* **2285**, Nonlinear Optical Properties of Organic Materials VII, San Diego, CA, pp. 366–375 (July 27–29, 1994).

[MKK88] T. Miyashita, M. Kawachi, and M. Kobayashi, "Silica-Based Planar Waveguides for Passive Components," THJ3, Conference on Optical Fiber Communications (1988).

[MKS93] M. Mori, Y. Kizaki, M. Saito, and A. Hongu, "A Fine Pitch COG Technique for a TFT-LCD Panel Using an Indium Alloy," *IEEE Proc 43rd Electronic Component and Tech Conf*, Orlando, FL (June 1–4, 1993).

[MKZ95] T. S. McLaren, S. Y. Kang, W. Zhang, D. Hellman, T. H. Ju, and Y. C. Lee, "Thermosonic Flip-Chip Bonding for an 8×8 VCSEL Array," *IEEE Proc 45th Electronics Components and Tech Conf*, Las Vegas, NV, pp. 393–400 (May 1995).

[MLF95] J. Ma, S. Lin, W. Feng, R. J. Feuerstein, and A. R. Mickelson, "Modeling Photobleached Optical Polymer Waveguides," *Appl Opt* **34**:24, p. 5352 (1995).

[MLM94] F. B. McCormick, A. L. Lentine, R. L. Morrison, J. M. Sasian, T. J. Cloonan, R. A. Novotny, M. G. Beckman, M. J. Wojcik, S. J. Hinterlong, and D. B. Buchholz, "155-Mb/s Operation of a FET-SEED Free-Space Switching Network," *IEEE Photonics Tech Lett* **6**:12, pp. 1479–1481 (1994).

[MLN96] V. Morozov, Y. C. Lee, J. Neff, D. O'Brien, T. McLaren, and H. Zhou, "Tolerance Analysis for Three-dimensional Optoelectronic Systems Packaging," *Opt Eng* **35**:7, pp. 2034–2044 (July1996).

[Mob94] K. Mobarhan, *InGaAsP/InP 1.3 μm Double Heterostructure Laser Grown on Si Substrate by Metalorganic Vapor Phase Epitaxy*, Ph.D. Dissertation, Northwestern University (December 1994).

[MoW95] R. A. Modavis and T. W. Webb, "Anamorphic Microlens for Laser Diode to Single-Mode Fiber Coupling," *IEEE Photonics Tech Lett* **7**:7 (July 1995).

[MPP95] The Second International Conference on Massively Parallel Processing Using Optical Interconnections, October 23–24, 1995, San Antonio, Texas, IEEE Computer Society, IEEE Technical Committee on Computer Architecture, IEEE, Piscataway, NJ.

[MST83] K.-I. Mizuisha, M. Sawai, S. Todoroki, S. Tsuji, M. Hirao, and M, Nakamura, "Reliability of InGaAsP/InP Buried Heterostructure 1.3 μm Lasers," *IEEE J Quant Electron* **QE-19**:8, pp. 1294–1301 (August 1983).

[MTN95] V. Morozov, H. Temkin, J. Neff, and A. Fedor, "Analysis of a 3-D Computer Optical Scheme Based on Bidirectional Free-Space Interconnects," *Opt Eng* **34**:2, pp. 523–534 (February 1995).

[Mun94] J. Mun, "Photodetectors and OEIC Receivers," in O. Wada, ed., *Optoelectronic Integration: Physics, Technology and Applications*, pp. 191–232 (Boston: Kluwer Academic Publishers, 1994).

[Muo95] T. V. Muoi, "Integrated Fiber Optic Transmitters and Receivers for SONET/ATM Applications," *ECTC 1995 Proc*, p. 1092 (1995).

[Mur88] E. J. Murphy, "Fiber Attachment for Guided Wave Devices," *J Lightwave Tech* **6**, pp. 862–871 (1988).

[MYY94] S. Mino, K. Yoshino, Y. Yamada, M. Yasu, and K. Moriwaki, "Optoelectronic Hybrid Integrated Laser Diode Module Using Silica-on-Terraced-Silicon Platform," *Proc IEEE Laser and Electrooptics Society Annual Meeting*, pp. 271–272 (1994).

[MZC89] A. Miliou, H. Zhenguang, H. C. Cheng, R. Srivastava, R. V. Ramaswamy, "Fiber-Compatible $K^+ -Na^+$ Ion Exchanged Channel Waveguides: Fabrication and Characterization," *IEEE J Quant Electr* **25**:8, p. 1889 (1989).

[Naj92] S. I. Najafi, ed., *Introduction to Glass Integrated Optics* (Boston: Artech House, 1992).

[Naj94] S. I. Najafi, ed., "Glass Integrated Optics and Optical Fiber Devices," Critical reviews of optical science and technology; V. CR53, Bellingham, WA: SPIE Optical Engineering Press, 1994.

[NBH93] R. A. Nordin, D. B. Buchholz, R. F. Huisman, N. R. Basavanhally, and A. F. J.

Levi, "High-Performance Optical Datalink Array Technology," *IEEE Trans Components, Hybrids and Manufacturing Tech* **16**:8, pp. 783–788 (1993).

[NBJ89] C. Nissim, A. Beguin, R. Jansen, and P. Laborde, Conference on Optical Fiber Communications (1989).

[NeC95] J. A. Neff and Y. Chen, "Miniature Optomechanical System Supporting Optical Interconnections Between Smart Pixel Arrays," Invention Disclosure, University of Colorado (June 1995).

[Nef91] J. A. Neff, "Massive Optical Interconnections for Computer Applications," *Proc SPIE* **1390** (1991).

[Nef94] J. A. Neff, "Optical Interconnects Based on Two-Dimensional VCSEL Arrays," *Proceedings of the First International Workshop on Massively Parallel Processing Using Optical Interconnections* (Piscataway, NJ: IEEE Computer Society Press, April 1994).

[Nef96] J. A. Neff, "Thermal Considerations of Free-Space Optical Interconnects Using VCSEL-Based Smart Pixel Arrays," *Proc SPIE* **2691** (February 1996).

[NeJ95] J. A. Neff and K. Johnson, "Optical Computers: Improving Medicine, Manufacturing, and the Military," *Photonics Spectra* **29**:11 (November 1995).

[Nem94] S. Nemoto, "Experimental Evaluation of a New Expression for the Far Field of a Diode Laser Beam," *Appl Opt* **33**:27, pp. 6387–6392 (September 20, 1994).

[NML96] J. A. Neff, V. Morozov, Y.-Ch. Lee, D. O'Brien, T. McLaren, and H. Zhou, "Tolerance Analysis for Three-Dimensional Optoelectronic Systems Packaging," *Opt Eng* **35**:7 (July 1996).

[Nor92] R. A. Nordin et al., "A Systems Perspective on Digital Interconnection Technology," *IEEE J Lightwave Tech* **10**:6, pp. 811–827 (June 1992).

[Nor93] R. A. Nordin et al., "High Performance Optical Data Link Array Technology," *Proc 43rd Electronic Components and Tech Conf* (IEEE), Orlando, FL, pp. 795–801 (June 1993).

[Nov93] R. A. Novotny et al., "Two-Dimensional Optical Data Lines," *Proc 43rd Electronic Components and Tech Conf* (IEEE), Orlando, FL, pp. 790–794 (June 1993).

[NYA91] S. Nagasawa, Y. Yokoyama, F. Ashiya, and T. Satake, "A Single-Mode Multifiber Push-On Connector with Low Insertion and High Return Losses," *Proc 17th Eur Conf on Opt Commun*, pp. 49–52 (1991).

[OIB96] SPIE 1996 Conference on Optical Interconnects in Broadband Switching Architectures, held at SPIE Photonics West '96, San Jose, California (January 27–February 2, 1996).

[Oik94] Y. Oikawa et al., "Packaging Technology for a 10-Gb/s Photoreceiver Module," *J Lightwave Tech* **12**:2, p. 343 (1994).

[OKO94] M. Okuno, K. Kato, Y. Ohmori, M. Kawachi, and T. Matsunaga, "Improved 8 × 8 Integrated Optical Matrix Switch Using Silica-Based Planar Lightwave Circuits," *IEEE J Lightwave Tech* **12**:9, pp. 1597–1606 (1994).

[Olb93] G. R. Olbright et al., "Micro-Optic and Microelectronic Integrated Packaging of Vertical Cavity Laser Arrays," Processing and Packaging of Semiconductor Lasers and Optoelectronic Devices, *Proc SPIE* **1851**, pp. 97–105 (1993).

[Opt96] Optivideo Corp., 5311 Western Avenue, Boulder, CO 80301.

[Pao75] T. L. Paoli, "A New Technique for Measuring the Thermal Resistance of Junction Lasers," *IEEE J Quantum Electron* **QE-11**:7, pp. 498–503 (July 1975).

[PaH83] D. R. Pape and L. J. Hornbech, "Characteristics of the Deformable Mirror Device for Optical Information Processing," *Opt Eng* **22**, pp. 675–681 (1983).

[PaH87] W. A. Payne and H. S. Hinton, "System Considerations for the Lithium Niobate Photonic Switching Technology," *Proc First Topical Meeting of Photonic Switching*, Incline Village, Nevada, pp. 196–199 (1987).

[PaL91] S. K. Patra and Y. C. Lee, "Modelling of Self-Alignment Mechanism in Flip-Chip Soldering Part II: Multichip Solder Joints," *Proc ECTC '91*, pp. 783–788 (1991).

[Pet82] K. E. Petersen, "Silicon as a Mechanical Material," *Proc IEEE* **70**:5, pp. 420–457 (1982).

[PIR96] Photonic Integration Research, Inc., 2727 Scioto Parkway, Columbus, OH 43221.

[Pol95] C. R. Pollock, *Fundamentals of Optoelectronics* (Burr Ridge, IL: Irwin, 1995).

[PPR93] K. Pedrotti, R. Pierson, C. Rarley, and M. Chang, "Monolithic Optical Integrated Receivers Using GaAs Heterojunction Bipolar Transistors," *Microwave J*, pp. 254–261 (May 1993).

[Pro94] G. M. Proudley et al., "Fabrication of Two-Dimensional Fiber Optic Arrays for an Optical Crossbar Switch," *Opt Eng* **33**:2, pp. 627–635 (1994).

[PSL92] S. K. Patra, S. S. Sritharan, and Y. C. Lee, "Minimum-Energy Surface Profile of Solder Joints for Non-Circular Pads," *ASME J Appl Mech*, pp. 390–397 (June 1995).

[Ref91] J. J. Refi, "Fiber Optic Cable: A Lightguide," ABC TeleTraining, Inc., pp. 23–25 (1991).

[Ros82] W. E. Ross et al., "Two-Dimensional Magneto-Optic Spatial Light Modulator for Signal Processing," *J Soc Photo-Optical Instr Eng* **341**, pp. 191–198 (1982).

[Sas94] J. M. Sasian et al., "Fabrication of Fiber Bundle Arrays for Free-Space Photonic Switching Systems," *Opt Eng* **33**:9 (1994).

[Sau92] J. R. Sauer et al., "Photonic Interconnects for Gigabit Multicomputer Communications," *IEEE Lightwave Telecommun Syst* **3**:3, pp. 12–19 (August 1992).

[SaY78] I. Sakama, Hiroo Yoneza et al., "Semiconductor Laser Device Equipped with a Silicon Heatsink," U.S. Patent No. 4,092,614 (May 30, 1978).

[SBC90] D. A. Smith, J. E. Baran, and K. W. Cheung, "Polarization-Independent Acoustically Tunable Optical Filter," *Appl Phys Lett* **56**:3, p. 209 (1990)

[SBR92] J. R. Sauer, D. J. Blumenthal, and A. V. Ramanan, "Photonic Interconnects for Gigabit Multicomputer Communications," *IEEE Lightwave Telecommun Syst Mag.*, pp. 12–19 (1992).

[SCA91] C. Shieh, J. Chi, C. Armiento, A. Negri, M. Rothman, and P. Haugsjaa, "A 1.3 μm Ridge Waveguide InGaAs P Laser on GaAs and Silicon Substrates by Thin-Film Transfer," *Proc Third Int Conf on InP and Related Compound Materials*, Cardiff, Wales, pp. 272–275 (April 8, 1991).

[SCF95] D. Schwartz, C. Chun, B. Foley, D. Hartman, M. Lebby, H. Lee, C. Shieh, S. Kuo, S. Shook, and B. Webb, "A Low-Cost, High-Performance Optical Interconnect," *Proc 45th Electronic Components and Tech Conf*, p. 376 (1995).

[Sch78] C. M. Schroeder, "Accurate Silicon Spacer Chips for an Optical Fiber Cable Connector," *Bell Syst Tech J* **57**, pp. 91–97 (1978).

[Sch95] D. B. Schwartz et al., "A Low-Cost, High-Performance Optical Interconnect," *Proc 45th Electronic Components and Tech Conf*, Las Vegas, NV, pp. 376–379 (1993).

[ScT91] E. H. Schmitt and D. B. Tuckerman, "Vacuum Die Attach for Integrated Circuits," U.S. Patent No. 5,046,656 (September 10, 1991).

[Sha90] V. S. Shah et al., "Efficient Power Coupling from a 980-nm, Broad-Area Laser to a Single-Mode Fiber Using a Wedge-Shaped Fiber Endface," *J Lightwave Tech* **8**:9 (September 1990).

[She85] G. Sheward, *High Temperature Brazing in Controlled Atmospheres* (Elmsford, NY: Pergamon Press, 1985), pp. 60–62.

[ShG79] S. K. Sheem and T. G. Giallorenzi, "Single-Mode Fiber Optical Power Divider: Encapsulated Etching Technique," *Opt Lett* **4**:1, pp. 29–31 (1979).

[Shi93] F. Shiomokawa et al., "A Fabrication Technology of Fluorinated Polyimide Waveguides on Copper-Polyimide Multilayer Substrate for Optoelectronic Multichip Module," *Proc 43nd ECTC*, pp. 705–710 (1993).

[SIA94] Semiconductor Industry Association, *The National Technology Roadmap for Semiconductors*, San Jose, CA: (Semiconductor Industry Association, 1994).

[Sim93] S. P. Sim, "The Reliability of Laser Diodes and Laser Transmitter Modules," *Microelect Rel* **33**:7 (1993).

[SLL94] O. T. Strand, M. E. Lowry, S. Y. Lu, D. C. Nelson, D. J. Nikkel, M. D. Pocha, and K. D. Young, "Automated Fiber Pigtailing Technology," *Proc 44th Electronic Components and Tech Conf*, pp. 1000–1003 (1994).

[SMN95] T. Sakano, T. Matsumoto, and K. Noguchi, "Three-Dimensional Board-to-Board Free-Space Optical Interconnects and their Application to the Prototype Multiprocessor System: COSINE-III," *Appl Opt* **34**:11 (April 10, 1995).

[SMS94] T. Sterling, P. C. Messina, and P. H. Smith, "Enabling Technologies for Peta(FL)OPS Computing," Caltech Concurrent Supercomputing Facilities Report #45 (June 1994).

[SnL83] A. W. Snyder and J. D. Love, *Optical Waveguide Theory* (New York: Chapman and Hall, 1983).

[Sto88] Q. F. Stout, "Mapping Vision Algorithms to Parallel Architectures," *Proc IEEE* **76**:8, pp. 982–995 (August 1988).

[SYA91] K. Shiraishi, T. Yanagi, Y. Aizawa, and S. Kawakami, "Fiber-Embedded In-Line Isolator," *J Lightwave Tech* **9**, p. 430 (1991).

[SzH95] T. H. Szymanski and H. S. Hinton, "Design of a Terabit Free-Space Photonic Backplane for Parallel Computing," *Proc Second Int Workshop on Massively Parallel Processing Using Optical Interconnections*, San Antonio, TX, pp. 16–27 (October 23–24, 1995).

[TAK92] A. Takai, H. Abe, and T. Kato, "Subsystem Optical Interconnections Using Long-Wavelength LD and Single-Mode Fiber Arrays," *Proc 42nd ECTC* (1992).

[Tak93] H. Takahara et al, "Optical Waveguide Interconnections for Optoelectronic Multichip Modules," *Proc SPIE* **1849**, Optoelectronic Interconnects, pp. 70–78 (January 18–20, 1993).

[Tak94] A. Takai et al., "200-Mb/s/ch 100-m Optical Subsystem Interconnections Using 8-Channel 1.3-μm Laser Diode Arrays and Single-Mode Fiber Arrays," *J Lightwave Tech* **12**:2, pp. 260–269 (February 1994).

[TaL96] Q. Tan and Y. C. Lee, "Soldering Technology for Optoelectronic Packaging,"

IEEE Proc 46th Electronic Components and Tech Conf, Orlando, FL (May 29–June 1, 1996).

[Tam90] T. Tamir, ed., *Guided-Wave Optoelectronics,* 2nd ed., Springer Series in Electronics and Photonics, Vol. 26 (New York: Springer-Verlag, 1990).

[THH94] F. Tian, Ch. Harizi, H. Herrmann, V. Reimann, R. Ricken, U. Rust, W. Sohler, F. Wehrmann, and S. Westenhofer, "Polarization-Independent Integrated Optical, Acoustically Tunable Double-Stage Wavelength Filter in $LiNbO_3$," *IEEE/OSA J Lightwave Tech* **12**:7, pp. 1192–1197 (1994).

[Thy88] L. Thylen, "Integrated Optics in $LiNbO_3$: Recent Developments in Devices for Telecommunications," *IEEE/OSA J Lightwave Tech* **6**:6, p. 847 (1988).

[TJY88] N. Takato, K. Jinguji, M. Yasu, H. Toba, and M. Kawachi, "Silica-Based Single-Mode Waveguides on Silicon and their Application to Guided-Wave Optical Interferometers," *IEEE/OSA J Lightwave Tech* **6**:6, pp. 1003–1010 (1988).

[TNM93] H. Tanaka, R. Nagaoka, T. Makabe, A. Kawatani, Y. Suzuki, Y. Kobayashi, and F. Suzuki, "A Multichip Module for the 156 Mbps Optical Interface," *Int J Microcircuits and Elect Packaging* **16**:4, pp. 293–300 (1993).

[ToC95] C. Tocci and H. J. Caulfield, eds., *Optical Interconnection: Foundations and Applications* (Boston: Artech House, 1995).

[TrN86] C. M. Truesdale and D. A. Nolan, "Multiple-Index Waveguide Coupler," Conference on Optical Fiber Communications, p. 60 (1986).

[TRW94] D. Tsang, H. Roussel, J. Woodhouse, J. Donnelly, C. Wang, D. Spears, R. Bailey, and D. Mull, "High-Speed, High-Density Parallel Free-Space Optical Interconnections," *Proc LEOS 1994 Annual Meeting,* pp. 217–216 (1994).

[Tsa95] D. Z. Tsang, "High-Density 300 Gbps/cm^2 Parallel Free-Space Optical Interconnection Design Considerations," *Optical Computing* **10**, 1995 OSA Technical Digest Series, pp. 277–279 (1995).

[TsS75] T. Tsukada and Y. Shima, "Thermal Characteristics of Buried-Heterostructure Injection Lasers," *IEEE J Quant Electron* **QE-11**:7, pp. 494–498 (July 1975).

[TuA93] T. A. Tumolillo, Jr., and P. R. Ashley, "Multilevel Registered Polymeric Mach-Zehnder Intensity Modulator Array," *Appl Phys Lett* **62**:24, p. 3068 (1993).

[TuR89] R. R. Tummala and E. J. Rymazewski, *Microelectronics Packaging Handbook* (New York: Van Nostrand Reinhold, 1986).

[TYK86] N. Takato, M. Yasu, and M. Kawatchi, "Low-Loss High-Silica, Single-Mode Channel Waveguides," *Electron Lett* **22**, pp. 321–322 (1986).

[USL93] K. S. Urquhart, R. Stein, and S. H. Lee, "Computer-Generated Holograms Fabricated by Direct Write of Positive Electron-Beam Resist," *Opt Lett* **18**:4, pp. 308–310 (February 15, 1993).

[UYH94] Y. Uematsu, Y. Yamabayashi, K. Hohkaawa, M. Togashi, N. Tanaka, and Y. Arai, "A 1.25-Gb/s Four-Channel GaAs MSI Integrated with MSM-PDs for Optical Interconnection," *Proc IEEE Laser and Electro-Optics Society Annual Meeting,* pp. 105–106 (1994).

[VBB91] P. Vettiger, M. Benedict, G. Bona, P. Buchmann, E. Cahoon, K. Datwyler, H. Dietrich, A. Moser, H. Seitz, O. Voegeli, D. Webb, and P. Wolf, "Full-Wafer Technology: A New Approach to Large-Scale Laser Fabrication and Integration," *IEEE J Quantum Elctron* **27**, pp. 1319–1331 (1991).

[VeS91] A. R. Vellekoop and M. K. Smit, "Four-channel Integrated Optic Wavelength Demultiplexer with Weak Polarization Dependence," *J Lightwave Tech* **9**, pp. 310–314 (1991).

[VOP92] P. Verboom, Y. S. Oei, E. Pennings, M. Smit, J. van Uffelen, H. van Brug, I. Moerman, G. Coudenijs, and P. Demeester, "A Long InGaAsP/InP Waveguide Section with Small Dimensions," *IEEE Photonics Tech Lett* **4**:10, p. 1112 (1992).

[VVS91] M. Varasi, A. Vannucci, M. Signorazzi, "Lithium Niobate Proton Exchange Technology for Phase-Amplitude Modulators," *Proc SPIE, Integrated Optical Circuits - V.1583*, pp. 165–169 (1991).

[Wad94] O. Wada, Ed., *Optoelectronic Integration: Physics, Technology and Applications* (Boston: Kluwer Academic Publishers, 1994).

[WaE90] M. J. Wale and C. Edge, "Self-Aligned, Flip-Chip Assembly of Photonic Devices with Electrical and Optical Connections," *IEEE Trans Components, Hybrids and Manuf Tech* **CHMT-13**:4, pp. 780–786 (1990).

[Wal89] M. J. Wale, "Self Aligned Flip Chip Assembly of Photonic Devices with Electrical and Optical Connections," *Proceedings ECTC '90*, p. 34 (1990); *Proceedings ECOC '89*, p. 368 (1989).

[Wan88] S. Y. Wang, "High-Speed Photodiodes and Electrooptic Waveguide Modulators," Conference on Optical Fiber Communications, p. 28 (1988).

[Wan92] C.-Y. Wang, *A New Bonding Technology Using Gold–Tin Multilayer Composites for Microelectronics and Photonics*, Ph.D. Dissertation, University of California at Irvine (1992).

[WCS90] D. Welch, R. Craig, W. Streifer, and D. Scrifres, "High Reliability, High Power, Single-Mode Laser Diodes," *Elect Lett* **26**:18 (August 30, 1990).

[Wel87] F. S. Welsh, "Lightwave Data Links and Interfaces," *AT&T Tech J* **66**:1, p. 65 (1987).

[Wil93] R. C. Williamson, *Proc Optical Computing Topical Meeting of the Optical Society of America*, Palm Springs, CA (March 16–19, 1993).

[WMB90] J. E. Watson, M. A. Milbrodt, K. Bahadori, M. F. Dautartas, C. T. Kemmerer, D. T. Moser, A. W. Schelling, T. O. Murphy, J. J. Veselka, and D. A. Herr, "A Low-Voltage 8×8 Ti : LiNbO$_3$ Switch with a Dilated-Benes Architecture," *IEEE/OSA J Lightwave Tech* **8**:5, p. 794 (1990).

[Won95] Y.-M. Wong et al., "Technology Development of a High-Density 32-Channel, 16-Gbps Optical Data Link for Optical Interconnection Applications for the Optoelectronic Technology Consortium (OETC)," *J Lightwave Tech* **13**:6, pp. 995–1016 (1995).

[YaK87] Y. Yamada and M. Kobayashi, "Single-Mode Optical Fiber Connection to High Silica Waveguide with Fiber Guiding Groove," *J Lightwave Tech* **5**, pp. 1716–1720 (1987).

[Yan93] M. Yanoet et al., "Skew-Free Parallel Optical Links and Their Array Technology," *Proc 45th Electronic Components and Tech Conf*, Las Vegas, NV, pp. 552–564 (1993).

[Yar91] A. Yariv, *Optical Electronics*, 4th ed., the Holt, Rinehart, and Winston Series in Electrical Engineering (Philadelphia: Saunders College Publication, 1991).

[YEG95] N. Yamanaka, K. Endo, K. Genda, H. Fukuda, T. Kishimoto, and S. Sasaki, "320-Gb/s High-Speed ATM Switching System Hardware Technologies Based on

Copper-Polyimide MCM," *IEEE Trans Comp, Packaging, Manuf Tech Part B* **18**:1, pp. 83–91 (1995).

[YeS94] H. J. Yeh and J. S. Smith, "Integration of GaAs Vertical-Cavity Surface-Emitting Lasers on Si by Substrate Removal," *Appl Phys Lett* **64**:12, pp. 1466–1468 (March 1994).

[YFL93] J. S. Yoo, S. Fang, and H. H. Lee, "Condition for No Thermal Runaway In CW Semiconductor Lasers," *J Appl Phys* **74**:11 (December 1, 1993).

[YKN95] M. Yamamoto, M. Kubo, and K. Nakao, "Si-OEIC with a Built-in Pin-Photodiode," *IEEE Trans Electr Dev* **42**, pp. 58–63 (1995).

[YMG94] D. R. Young, A. J. Morrow, K. Gadkaree, and D. E. Quinn, "Packaging of High Reliability Couplers," *Proc ECTC '94*, p. 1004 (1994).

[YMT92] K. Yamamoto, K. Mizuuchi, and T. Taniuchi, "Low-Loss Channel Waveguides in MgO:LiNbO$_3$ and LiTaO$_3$ by Pyrophosphoric Acid Proton Exchange," *Jpn J Appl Phys Part 1* **31**:4, p. 1059 (1992).

[YOK90] N. Yamanaka, T. Ohsaki, S. Kikuchi, and T. Kon, "1.8 Gb/s High-Speed Space Division Switching Module using Copper Polyimide Multilayer Substrate," *Proc 40th ECTC*, pp. 562–570 (1990).

[YSD93] S. L. Yellen, A. H. Shepard, R. L. Dalby, J. A. Baumann, H. B. Serreze, T. S. Guido, R. Soltz, K. J. Bystrom, C. M. Harding, and R. G. Waters, "Reliability of GaAs-Based Semiconductor Diode Lasers: 0.6–1.1 m," *IEEE J Quantum Electronics* **29**:6 (June 1993).

[YTO93] Y. Yamada, A. Takagi, I. Ogawa, M. Kawachi, and M. Kobayashi, "Silica-Based Optical Waveguide on Terraced Silicon Substrates as Hybrid Integration Platform," *Electronics Lett* **29**, pp. 444–446 (1993).

[YUK79] H. Yonezu, M. Ueno, T. Kamejima, and I. Hayashi, "An AlGaAs Window Structure Laser," *IEEE J Quantum Electronics* **15**:8 (August 1979).

[ZMS94] J. Z. Zhang, D. T. McAvoy, and K. Schaschek, "Jet Vapor Deposition of Organic Guest-Inorganic Host Thin Films for Optical and Electronic Applications," *J Electron Mater* **23**:11, p. 1239 (1994).

[ZHK94] T. Zyung, W.-Y. Hwang, and J.-J. Kim, "Accelerated Photobleaching of Nonlinear Optical Polymer for the Formation of Optical Waveguide," *Appl Phys Lett* **64**:26, p. 3527 (1994).

[ZMN96] H. Zhou, V. Morozov, J. Neff, and A. Fedor, "Analysis of VCSEL-Based Free-Space Bidirectional Interconnects," *Appl Opt* (1996), submitted.

[Zmu94] C. A. Zmudzinski et al., "Uniform Near-Field, Flat-Phasefront Antiguided Power Amplifier with Three-Core ARROW Master Oscillator," *CLEO 1994 Tech Digest* **8**, paper CMA2 (1994).

[Zor94] P. Zorabedian, "Axial-Mode Instability in Tunable External-Cavity Semiconductor Lasers," *J Quantum Electr* **30**:7 (July 1994).

Index

巴楚

（近台北火車站）

Tel: (02) 2331-0940

(02) 2383-2952

Fax: (02) 2361-3007

850

台北市懷寧街

3 6 號 2 樓

林彥政